SIGNALLING NETWORKS IN CELL SHAPE AND MOTILITY

The Novartis Foundation is an international scientific and educational charity (UK Registered Charity No. 313574). Known until September 1997 as the Ciba Foundation, it was established in 1947 by the CIBA company of Basle, which merged with Sandoz in 1996, to form Novartis. The Foundation operates independently in London under English trust law. It was formally opened on 22 June 1949.

The Foundation promotes the study and general knowledge of science and in particular encourages international co-operation in scientific research. To this end, it organizes internationally acclaimed meetings (typically eight symposia and allied open meetings and 15–20 discussion meetings each year) and publishes eight books per year featuring the presented papers and discussions from the symposia. Although primarily an operational rather than a grant-making foundation, it awards bursaries to young scientists to attend the symposia and afterwards work with one of the other participants.

The Foundation's headquarters at 41 Portland Place, London W1B 1BN, provide library facilities, open to graduates in science and allied disciplines. Media relations are fostered by regular press conferences and by articles prepared by the Foundation's Science Writer in Residence. The Foundation offers accommodation and meeting facilities to visiting scientists and their societies.

Information on all Foundation activities can be found at http://www.novartisfound.org.uk

The Institute of Molecular and Cell Biology (IMCB) is a member of the Agency for Science, Technology and Research (A*STAR). Established in 1987, the research institute's mission is to foster a vibrant research culture for biomedical sciences and high quality manpower training to facilitate development of the biotechnology and pharmaceutical industries in Singapore.

Funded primarily by Biomedical Research Council (BMRC) of A*STAR, IMCB has about 35 core research labs and 8 core facility units consisting of over 400 research scientists. IMCB's research activities focus on five major fields: **Cell Biology**, **Developmental Biology**, **Structural Biology**, **Infectious Diseases** and **Cancer Biology**. IMCB continues to publish in renowned international journals, with more than a 1000 publications since 1987.

IMCB is currently based at The Biopolis @ One North. It is envisioned to be the biggest Biomedical Sciences R&D hub in Asia. IMCB continues to strive for excellence in biomedical R&D and the vision of Singapore as a world class hub for the Biomedical Sciences in Asia and beyond.

Novartis Foundation Symposium 269

SIGNALLING NETWORKS IN CELL SHAPE AND MOTILITY

2005

John Wiley & Sons, Ltd

Published in 2005 by John Wiley & Sons Ltd,
The Atrium, Southern Gate,
Chichester PO19 8SQ, UK

National 01243 779777
International (+44) 1243 779777
e-mail (for orders and customer service enquiries): cs-books@wiley.co.uk
Visit our Home Page on http://www.wileyeurope.com
or http://www.wiley.com

Other Wiley Editorial Offices

John Wiley & Sons Inc., 111 River Street, Hoboken, NJ 07030, USA

Jossey-Bass, 989 Market Street, San Francisco, CA 94103-1741, USA

Wiley-VCH Verlag GmbH, Boschstr. 12, D-69469 Weinheim, Germany

John Wiley & Sons Australia Ltd, 33 Park Road, Milton, Queensland 4064, Australia

John Wiley & Sons (Asia) Pte Ltd, 2 Clementi Loop #02-01, Jin Xing Distripark, Singapore 129809

John Wiley & Sons Canada Ltd, 22 Worcester Road, Etobicoke, Ontario, Canada M9W 1L1

Wiley also publishes its books in a variety of electronic formats. Some content that appears in print may not be available in electronic books.

Novartis Foundation Symposium 269
ix+239 pages, 34 figures, 0 tables

British Library Cataloguing in Publication Data

A catalogue record for this book is available from the British Library

ISBN-13 978-0-470-01190-4 (HB)
ISBN-10 0-470-01190-4 (HB)

Typeset in 10½ on 12½ pt Garamond by Dobbie Typesetting Limited, Tavistock, Devon.
Printed and bound in Great Britain by T. J. International Ltd, Padstow, Cornwall.
This book is printed on acid-free paper responsibly manufactured from sustainable forestry, in which at least two trees are planted for each one used for paper production.

Contents

Novartis Foundation symposium on Signalling networks in cell shape and motility held in collaboration with the Institute of Cell and Molecular Biology, Singapore, in Singapore on August 30th–1st September 2004

Editors: Gregory Bock (Organizer) and Jamie Goode

This symposium is based on a proposal made by Uttam Surana

Gary Borisy Chair's introduction 1

Tadaomi Takenawa From N-WASP to WAVE: key molecules for regulation of cortical actin organization 3
Discussion 10

Margaret A. Titus A conserved role for myosin VII in adhesion 16
Discussion 24

General discussion I 30

David G. Drubin, Marko Kaksonen, Christopher Toret and **Yidi Sun** Cytoskeletal networks and pathways involved in endocytosis 35
Discussion 43

Michele Knaus, Philippe Wiget, Yukiko Shimada and **Matthias Peter** Control of cell polarity in response to intra- and extracellular signals in budding yeast 47
Discussion 54

Fred Chang, Becket Feierbach and **Sophie Martin** Regulation of actin assembly by microtubules in fission yeast cell polarity 59
Discussion 66

Atsuo T. Sasaki and **Richard A. Firtel** Finding the way: directional sensing and cell polarization through Ras signalling 73
Discussion 87

Takashi Watanabe, Jun Noritake and **Kozo Kaibuchi** Roles of IQGAP1 in cell polarization and migration 92
Discussion 101

Gregg G. Gundersen, Ying Wen, Christina H. Eng, Jan Schmoranzer, Noemi Cabrera-Poch, Edward J. S. Morris, Michael Chen and **Edgar R. Gomes** Regulation of microtubules by Rho GTPases in migrating cells 106
Discussion 116

Eyal D. Schejter Actin organization in the early *Drosophila* embryo 127
Discussion 138

Yasuyuki Fujita and **Vania Braga** Epithelial cell shape and Rho small GTPases 144
Discussion 155

W. James Nelson, Frauke Drees and **Soichiro Yamada** Interaction of cadherin with the actin cytoskeleton 159
Discussion 168

Martin J. Humphries, Zohreh Mostafavi-Pour, Mark R. Morgan, Nicholas O. Deakin, Anthea J. Messent and **Mark D. Bass** Integrin–syndecan co-operation governs the assembly of signalling complexes during cell spreading 178
Discussion 188

Keith Mostov, Paul Brakeman, Anirban Datta, Ama Gassama, Leonid Katz, Minji Kim, Pascale Leroy, Max Levin, Kathleen Liu, Fernando Martin, Lucy E. O'Brien, Marcel Verges, Tao Su, Kitty Tang, Naoki Tanimizu, Toshiyuki Yamaji and **Wei Yu** Formation of multicellular epithelial structures 193
Discussion 200

Kathryn M. Eisenmann, Jun Peng, Bradley J. Wallar and Arthur S. Alberts Rho GTPase–formin pairs in cytoskeletal remodelling 206
Discussion 219

Final discussion 223

Index of contributors 231

Subject index 233

Participants

Art Alberts Laboratory of Cell Structure and Signal Integration, Van Andel Research Institute, 333 Bostwick Avenue NE, Grand Rapids, MI 49503, USA

Mohan Balasubramanian Temasek Life Sciences Laboratorories, 1 Research Link, Singapore 117604

Gary Borisy *(Chair)* Department of Cell and Molecular Biology, Northwestern Medical School, 303 E. Chicago Ave., Chicago, IL 60611, USA

Vania Braga Cell & Molecular Biology Section, Division of Biomedical Sciences, Faculty of Medicine, Imperial College, Sir Alexander Fleming Building, London SW7 2AZ, UK

Mingjie Cai Institute of Molecular and Cell Biology, Proteos, 61 Biopolis Drive, Singapore 138673

Fred Chang Department of Microbiology, Columbia University, 701 168th St. Rm 1404, New York, NY 10032, USA

David Drubin University of California, Berkeley, Department of Molecular & Cell Biology, 16 Barker Hall #3202, Berkeley, CA 94720-3202, USA

Richard Firtel Center for Molecular Genetics, 0380, University of California, San Diego, 9500 Gilman Drive, La Jolla, CA 92093-0380, USA

Feng Gu Novartis Institute for Tropical Diseases, 10 Biopolis Road, Chromos #05-01, Singapore 138670

Gregg Gundersen Department of Anatomy and Cell Biology, Columbia University, 1217 Black Building, 630 West 168th Street, New York, NY 10032, USA

Nick Harden Department of Molecular Biology and Biochemistry, Simon Fraser University, Office SSB 7146, 8888 University Drive, Burnaby, British Columbia V5A 1S6, Canada

Wanjin Hong Institute of Molecular and Cell Biology, Proteos, 61 Biopolis Drive, Singapore 138673

Martin Humphries School of Biological Sciences, University of Manchester, Michael Smith Building, Oxford Road, Manchester M13 9PT, UK

Kozo Kaibuchi Department of Cell Pharmacology, Nagoya University Graduate School of Medicine, 65 Tsurumai, Showa, Nagoya, Aichi 466-8550, Japan

Birgit Lane Division of Cell and Developmental Biology, Dundee University School of Life Sciences, MSI/WTB Complex, Dow Street, Dundee DD1 5EH, UK

David Lane Department of Surgery & Molecular Oncology, University of Dundee, Ninewells Hospital & Medical School, Dundee DD1 9SY, UK

Louis Lim Institute of Molecular and Cell Biology, Proteos, 61 Biopolis Drive, Singapore 138673

Frank Luca School of Veterinary Medicine, 3800 Spruce Street, Room 154E, Philadelphia, PA 19104, USA

Edward Manser Institute of Molecular and Cell Biology, Proteos, 61 Biopolis Drive, Singapore 138673

Keith Mostov Genentech Hall, Room N212B, Box 2140, 600 16th Street, San Francisco, CA 94143-2140, USA

W. James Nelson Department of Molecular & Cellular Physiology, Stanford University School of Medicine, Beckman Center, 279 Campus Drive, Stanford, CA 94305-5345, USA

Matthias Peter Institute of Biochemistry, HPM G 8.0, ETH Hönggerberg, 8093 Zürich, Switzerland

Eyal Schejter Department of Molecular Genetics, Belfer Building, Weizmann Institute of Science, Rehovot 76100, Israel

Michael Sheetz Columbia University, Biological Sciences, 600 Sherman Fairchild Centre, 1212 Amsterdam Avenue, New York, NY 10027, USA

Uttam Surana Institute of Molecular and Cell Biology, Proteos, 61 Biopolis Drive, Singapore 138673

Tadaomi Takenawa Department of Biochemistry, Institute of Medical Science, University of Tokyo, 4-6-1 Shirokanedai, Mianato-ku, Tokyo 108-8639, Japan

Margaret A. Titus Department of Genetics, Cell Biology, and Development, University of Minnesota, 6-160 Jackson Hall, 321 Church St. SE, Minneapolis, MN 55455, USA

Richard Vallee Columbia University, College of Physicians and Surgeons P45 15-409, Department of Pathology, 630 West 168th Street, New York, NY10032, USA

Alpha Yap Institute for Molecular Bioscience, University of Queensland, St Lucia, QLD 4072, Australia

Chair's introduction

Gary Borisy

Department of Cell and Molecular Biology, Northwestern Medical School, 303 E. Chicago Ave., Chicago, IL 60611, USA

Our task at this meeting is to talk about signalling networks in cell shape and motility. Signalling has always scared me because I don't understand it. But of course cells do respond to signals in some general sense. Signals from the extracellular environment, other cells, cell contact, substrates and even from inside the cell itself affect the cell. They are detected by sensors at the cell membrane (if we are talking about extracellular signals) and they go through this signalling network, and for the purposes of our discussion they arrive at the cytoskeleton. We will have a lot of animated discussion to try to direct us through this maze, and help us figure out how signalling networks accomplish changes in the cytoskeleton. This makes me think of going into a restaurant. I give a signal: I sit down at the table and give a signal to the waiter that I would like the ham and cheese sandwich. The waiter transmits this signal to someone else, and eventually the chef gets it. What we are particularly interested in is what goes on in the kitchen: who actually prepares the sandwich? I think of signalling as the waiter and the sequence getting the order to the kitchen, but what really is important to us is the preparation of the food. Who is implementing the signals, and how are they implemented? What does it mean to detect and respond to a signal? This will be one of the questions we will be addressing over the next few days.

In general terms the topics for our discussion have a long lineage. D'Arcy Thompson talked about cell form and wrote a nice book on the subject (Thompson 1917). He tried to understand cell form in terms of physical principles. Now we are looking inside at molecules and how they organize to generate cell form. It is a deep question. There are two major ways in which we want to discuss the signalling network and its interaction with the cytoskeleton: how do cells polarize, and how do they move directionally? Let's use the metaphor of a car for looking at how cells respond to external signals. What are the external signals? Traffic lights, the stop signs, road markings. But there has to be something inside the car — a driver — with an apparatus for detecting those signals and then responding to them. Underneath this level is the machinery itself. A car with no driver but an engine that is on will move, but it will go in more or less the direction the wheels were initially set in. The machinery of a car is intrinsically polarized. But it needs a driver to control the motility, and the whole system does respond to external cues. Are there equivalents to these levels of organization in cells? We don't have any driver who is conscious of the signals: it is all self-organizing. We have a self-organizing driver and a self-organizing machinery. What is the meaning of 'driver' and 'machinery' and detection of signals when it all has to be self organized?

I want to highlight this with one example from my own work. We take a little piece of cytoplasm which has been induced to be released from a cell by treatment. These fragments contain no nucleus, centrosome or microtubules. The fragments are active and ruffle around the edges. If perturbed they will move persistently in some direction. Without the perturbation they are symmetrical. With mechanical perturbation the symmetry is broken and persistent motility is achieved. The cell must have some internal machinery that is capable of persistent motion, like a car without a driver. How does this self-organize or self-propagate? How is the polarity maintained? These are some of the questions that we will be addressing over the next few days.

- How do cells become polarized?
- How do cells break symmetry and maintain that polarity?
- How do cells sense direction?
- How do they amplify shallow gradients of attractants or repellents?
- How do they detect the substrate and cell contacts?
- How do they respond to their own intracellular organization?

Several speakers will be talking about motility, including lamellipodia, motors, adhesion and actin dynamics. Many of the speakers are focusing on questions related to polarity, not just in terms of motion but also in terms of growth or organization of epithelial cells. When we come to the internal organization of cytoplasm, we will see that the actin organization is very important, and so is the microtubule network. The cross-talk between these two will be featured.

I'll end this introduction by saying that I hope we have an active and vigorous discussion over the next three days.

Reference

Thompson DW 1917 On growth and form. Cambridge University Press, Cambridge

From N-WASP to WAVE: key molecules for regulation of cortical actin organization

Tadaomi Takenawa

Institute of Medical Science, University of Tokyo and CREST, Japan Science and Technology Corporation (JST), Shirokanedai, Minato-ku 108-8639 Tokyo, Japan

Abstract. We first isolated N-WASP as one of the proteins bound to Ash/Grb2 SH3 domain. This protein has a VCA region (verplorin-like, cofilin-like, acidic region) at the C-terminus, which binds to G-actin and Arp2/3 complex, and several functional domains at the N-terminus, such as WHD (WASP homology domain) and GBD/CRIB domain. N-WASP activates Arp2/3 complex-dependent actin polymerization through the VCA region, leading to filopodium formation. Next, we found WAVE1, WAVE2 and WAVE3. All these proteins have also VCA regions at C-terminal areas and induce membrane ruffle formation. To clarify the different roles of WAVE1 and WAVE2, we established WAVE1- and WAVE2-deficient mouse embryonic fibroblasts (MEFs), because these two WAVEs are expressed in MEF. When wild-type MEFs are stimulated randomly by PDGF, two types of ruffles, peripheral and dorsal, are formed. However, dorsal ruffle formation does not occur in WAVE1-deficient MEFs. In contrast, peripheral ruffle formation is diminished in WAVE2-deficient MEFs. On the other hand, in MEFs migrating towards a chemoattractant gradient, only peripheral ruffles (lamellipodia) are formed. In this migration, WAVE1-deficient MEFs still could form lamellipodia but WAVE2-deficient MEFs could not. All these data show that WAVE2 but not WAVE1 is essential for lamellipodium formation and directed migration.

2005 Signalling networks in cell shape and motility. Wiley, Chichester (Novartis Foundation Symposium 269) p 3–15

WASP and WAVE: the background

More than 10 years have passed since we first identified a novel adaptor protein with one SH2 domain and two SH3 domains that we named Ash (abundant SH) (Matuoka et al 1992). Ash transmits upstream tyrosine kinase signals to Sos, a GDP/GTP exchange factor of Ras, leading to Ras activation. At that same time, Lowenstein et al (1992) identified the same adaptor protein and gave it the more popular name Grb2. It is now known that Ash/Grb2 plays crucial roles in activating Ras-Raf-MAPK pathways in response to a variety of growth factors (Schlessinger 2000). We previously observed that microinjection of anti-Ash/Grb2 antibody into fibroblasts inhibited both epidermal growth factor (EGF)-stimulated DNA synthesis and cytoskeletal reorganization (Matuoka et al 1993). These results suggested that a molecule downstream of Ash/Grb2 is involved in organization of the cytoskeleton. On the basis of these findings, we initiated studies of how the cytoskeleton is regulated downstream of extracellular stimuli, and we identified the N-WASP (neural Wiskott–Aldrich syndrome protein)/WAVE (WASP family verprolin-homologous protein) family of proteins. We first screened for a protein that transmits tyrosine kinase signals to molecules involved in cytoskeletal organization by affinity chromatography. Affinity chromatography was done with GST–Ash/Grb2 and later with phosphotyrosine-containing peptide having the Ash/Grb2 SH2 domain binding sequence. We loaded bovine brain lysates onto these columns and isolated 6 (180 kDa, 150 kDa, 120 kDa, 100 kDa, 65 kDa and 55 kDa) Ash/Grb2-binding proteins (Miura et al 1996). Partial amino acid sequences revealed that the 180 kDa, 120 kDa, 100 kDa and 55 kDa proteins were SOS, C-Cbl, dynamin and β-tubulin, respectively. In contrast, the 150 kDa and 65 kDa proteins appeared to be novel proteins, and we isolated the respective cDNAs. We now know that the 150 kDa protein is synaptojanin (Sakisaka et al 1997) and the 65 kDa protein is N-WASP.

N-WASP (Miki et al 1996) shows 50% amino acid identity to Wiskott–Aldrich syndrome protein (WASP) and is ubiquitous but enriched in neural tissues, whereas expression of WASP is restricted to haematopoietic cells. Because it was known that a defect in WASP function induced eczema, bleeding and recurrent infection, and altered the architecture of the cytoskeleton in T cells (Derry et al 1994), we hypothesized that N-WASP is involved in reorganization of the cytoskeleton. N-WASP can be manipulated more easily than WASP because purified WASP and exogenously expressed WASP in cells tend to form aggregates. Therefore, most studies have been done with N-WASP rather than WASP.

N-WASP is activated downstream of Cdc42 and phosphatidylinositol 4,5-bisphosphate (PIP2)

N-WASP contains a Cdc42-binding site, the GBD/CRIB motif, suggesting that N-WASP is a downstream target of Cdc42. Indeed, co-expression of N-WASP and active Cdc42 induces formation of long filopodia (Miki et al 1998a). A mutant of N-WASP that cannot bind Cdc42 does not induce filopodium formation even in the presence of Cdc42, indicating that N-WASP is necessary for filopodium formation downstream of Cdc42. There is a VCA (verprolin-cofilin-acidic) region at the C-terminal area of N-WASP. The V region recruits actin monomer

(Miki & Takenawa 1998) and the CA region binds to the Arp2/3 complex. We next showed that the VCA region is the minimal essential region necessary for Arp2/3-induced actin polymerization (Rohatgi et al 1999). However, full-length N-WASP does not activate Arp2/3 complex-dependent actin polymerization as strongly as the VCA fragment, suggesting that the VCA is masked through an intramolecular interaction. Indeed, an interaction between the GBD/CRIB domain and the VCA region (autoinhibited conformation) prevents binding of the Arp2/3 complex to the CA region. This autoinhibited conformation is released by the addition of Cdc42 and phosphatidylinositol-4,5-bisphosphate (PIP2) (Rohatgi et al 1999). Thus, Cdc42 and PIP2 release the autoinhibition of N-WASP and then activate the Arp2/3 complex, resulting in actin polymerization.

A big WAVE is coming

We learned from our extensive studies of N-WASP that the VCA region plays a crucial role in Arp2/3 complex activation. Thus, we next screened for novel VCA-containing proteins in DNA databases and found that KIAA0269 contains a VCA region (Miki et al 1998b). We named this protein WAVE1. At the same time, a *Dictyostelium* homologue of WAVE was identified and named Scar (Bear et al 1998). We isolated three WAVEs, WAVE1, -2, and -3 (Suetsugu et al 1999). These proteins are highly homologous to each other, and all contain a WAVE/Scar homology domain (WHD) at the N-terminus, a basic region, a proline-rich region, and a VCA region at the C-terminus. Therefore, the C-terminal regions of WAVEs are highly homologous to C-terminal region of N-WASP. There are presently five known VCA-containing proteins (WASPs and WAVEs) in mammals (Takenawa & Miki 2001). The structures and effector molecules of the WASPs and WAVEs are shown in Fig. 1. However, all WAVEs do not contain a Cdc42-binding site or a GBD/CRIB motif, suggesting that WAVEs are regulated differently than WASPs. Yeast two-hybrid system analyses yielded an adapter protein, IRSp53, that associates with WAVE2 (Miki et al 2000). IRSp53 binds through the SH3 domain present in its C-terminus to the proline-rich region of WAVE2. IRSp53 also associates with activated Rac through a Rac-binding domain in the N-terminus, linking the Rac signal to Arp2/3 complex-mediated actin polymerization and formation of lamellipodia (Fig. 2). Curiously, IRSp53 binds predominantly to WAVE2, and binding to WAVE1 and WAVE3 is very weak. This pathway presumably plays a significant role in several kinds of cells because a dominant-negative (ΔSH3) IRSp53 inhibits Rac-induced formation of lamellipodia. However, Eden et al (2002) recently proposed a novel mechanism for activation of WAVE1. They isolated protein complexes containing WAVE1 from bovine brain and identified the protein components by mass spectrometry. The isolated complex contained WAVE1, PIR121/Sra1, Nap1/Kette, Abi and

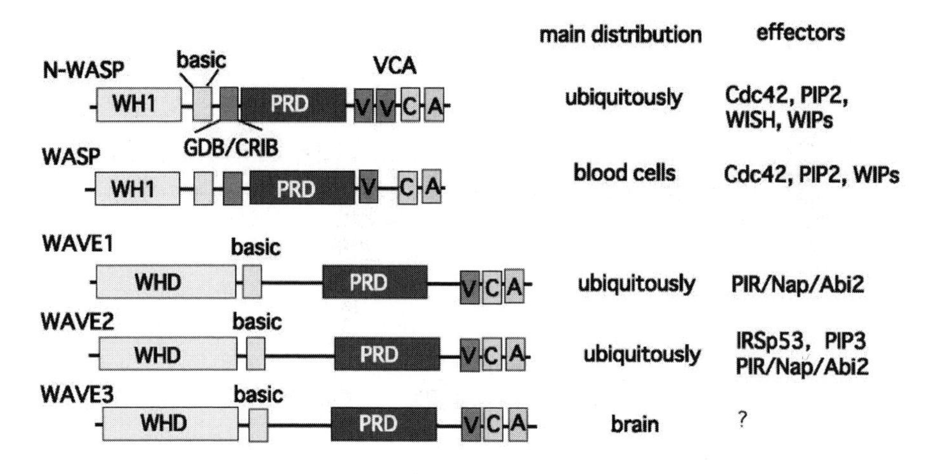

FIG. 1. Structures of WASP/WAVE family proteins and upstream and downstream effector molecules.

HSPC300. In this complex, activity of WAVE1 was suppressed. However, active Rac or Nck released the inhibition by causing dissociation of WAVE1 and HSCP300 from the complex. This dissociation activates WAVE1, which is thought to be constitutively active when present as a monomer, resulting in actin nucleation through Arp2/3 complex activation (Fig. 2). This model is very attractive, but it is inconsistent with recent findings. Abi1 was also identified as an essential component of the PIR121-Nap1-WAVE2 complex, and Abi was found to bind directly to WAVE2 through its WHD domain. Furthermore, WAVE1-Abi1-Nap1-PIR121 and WAVE2-Abi1-Nap1-PIR121 complex as reconstituted *in vitro* were as active as WAVE-Abi1 complex for Arp2/3 complex-dependent actin polymerization (Innocenti et al 2004, Stradal et al 2004). It was also reported that WAVE2 does not dissociate from PIR121 and Nap1, even in the presence of excess amounts of GTP-Rac, suggesting that the WAVE2-Abi1-Nap1-PIR121 complex remains intact under Rac activation. Blockade of expression of WAVE, Abi1, Nap1 or PIR121 by RNAi suppressed Rac-induced formation of membrane ruffles, indicating that Abi1, Nap1, and PIR121 are positive regulators of the cytoskeleton. Interestingly, recent RNAi studies showed that degradation of WAVE proteins was increased in Abi1-, Nap1- or PIR121-knockdown cells. Thus, it is possible that formation of complexes, containing WAVE is more important in stabilization of WAVE than in regulation of WAVE activity. Several activation pathways may be acting under different signalling pathways.

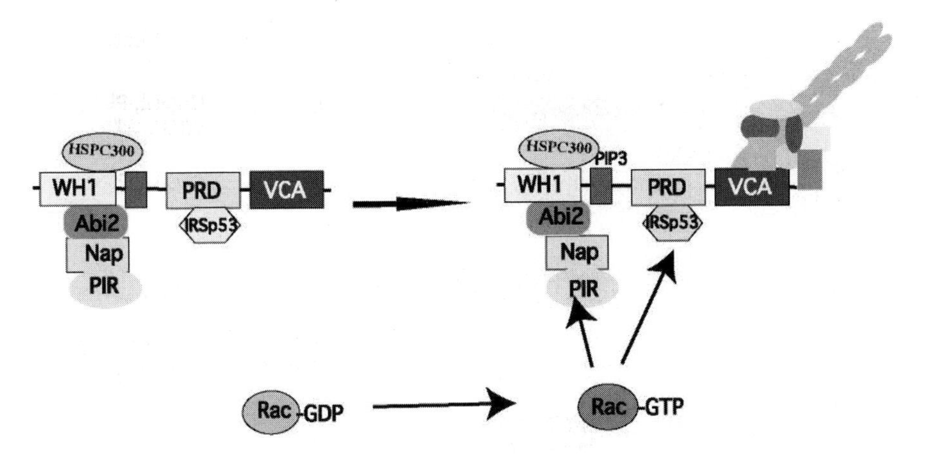

FIG. 2. Mechanisms of activation of WAVEs. Proposed mechanism 1: WAVE forms a complex with Abi2, Nap1 and PIR121. Formation of this complex stabilizes the WAVEs, thereby reducing degradation of WAVE proteins. Abi2 may recruit WAVEs to membrane ruffling sites. Proposed mechanism 2: WAVE forms a complex with IRSp53. When cells are stimulated and Rac is activated, the conformation of IRSp53 is changed, leading to activation of WAVEs. As a result, Arp2/3 complex is activated, leading to membrane ruffle formation.

Differential roles of WAVE1 and WAVE2 in membrane ruffling and directed migration

Two types of ruffles (peripheral ruffles and dorsal ruffles) are induced in mouse embryonic fibroblasts (MEFs) by platelet-derived growth factor (PDGF) stimulation (Fig. 3a). WAVE1 is present predominantly in dorsal ruffles, and WAVE2 is localized in peripheral ruffles (Fig. 3a), suggesting that WAVE1 and WAVE2 have different functions. To investigate their specific functions, we established WAVE1- and WAVE2-deficient MEFs from WAVE1 and WAVE2 knockout mice (Suetsugu et al 2003). Using these two cell lines, we studied the differential roles in ruffle formation and directed cell migration. When wild-type MEFs were stimulated randomly with PDGF, peripheral ruffles formed first, and dorsal ruffles formed after a brief time lag. In WAVE1-deficient MEFs, formation of dorsal ruffles did not occur in response to PDGF, whereas peripheral ruffle formation was not affected. On the contrary, in WAVE2-deficient MEFs, peripheral ruffle formation disappeared, whereas formation of dorsal ruffles was still induced by PDGF. These results indicate that WAVE1 regulates dorsal ruffle formation, and WAVE2 controls peripheral ruffle formation.

We next studied the roles of WAVE1 and WAVE2 in directed migration toward chemoattractant gradients (Fig. 3b). Directed migration of MEFs was tested with

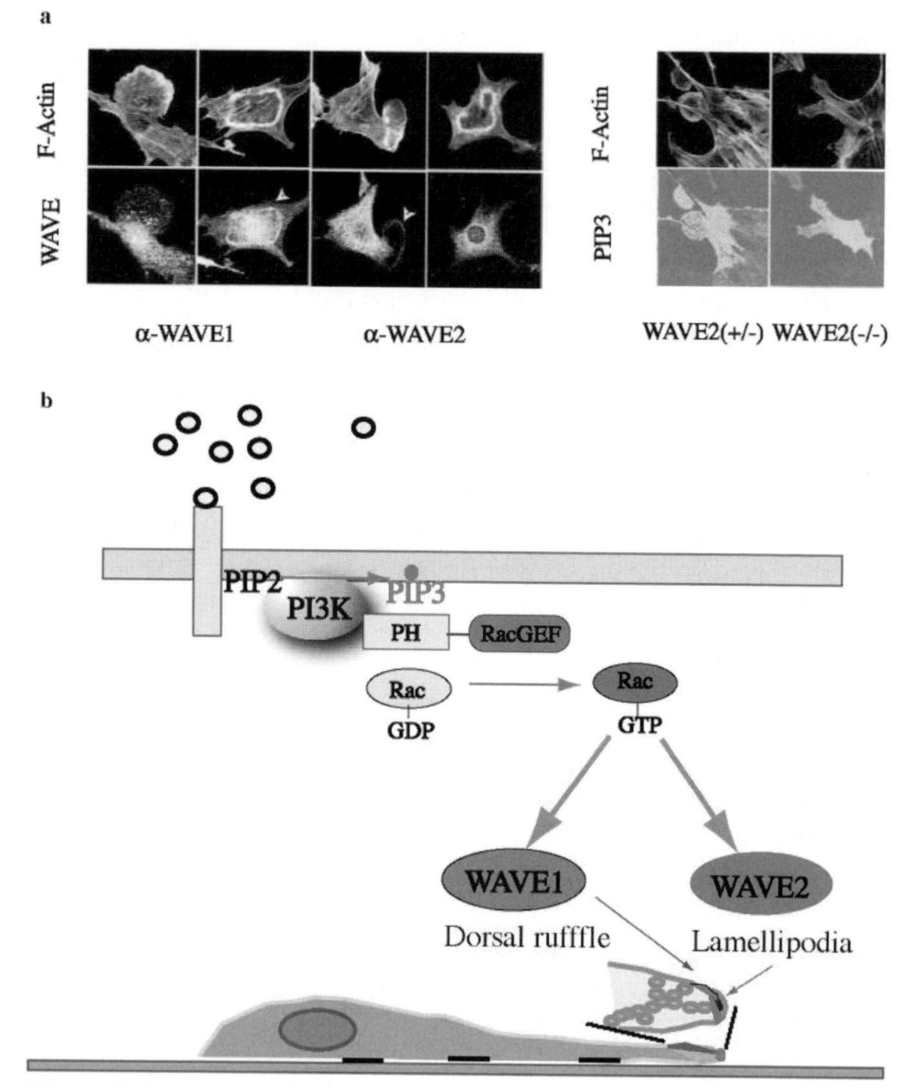

FIG. 3. Different roles of WAVE1 and WAVE2 in membrane ruffle formation and directed migration. (a) Localization of WAVE1 and WAVE2. When MEFs are stimulated with PDGF, two types of ruffles, dorsal and peripheral ruffles, are formed. WAVE1 localizes predominantly in dorsal ruffles, and WAVE2 localizes in peripheral ruffles. (b) Proposed signalling pathways for directed migration. When cells are stimulated with chemoattractants, cell polarity is established and a leading edge facing chemoattractant gradient is formed. PI3K is first activated at the place facing the highest chemoattractant concentrations, where PIP3 is produced. Rac-GEFs, which have PH domains, are then activated, leading to accumulation of the GTP form (active form) of Rac. Rac in turn activates WAVEs, resulting in Arp2/3 complex activation at the leading edge. Finally, mesh-like actin filaments that generate the movement force toward chemoattractants are assembled.

PDGF-coated heparin beads as chemoattractants. In this system, MEFs formed a leading edge facing a chemoattractant gradient, where peripheral ruffles, so-called lamellipodia, would also be expected to form. In cells migrating toward chemoattractants, dorsal ruffle formation was not observed. In this case, WAVE1 localized at the upper areas of lamellipodia instead of the dorsal ruffles, and WAVE2 was present at the front of lamellipodia. Lamellipodium formation facing PDGF-coated beads did not occur in WAVE2-deficient MEFs, although WAVE1-deficient MEFs still produced lamellipodia. Thus, WAVE2 but not WAVE1 is essential for lamellipodium formation and directed migration. When cells are exposed to chemoattractant gradients, cell polarity is first established to determine the direction of migration. This cell polarity is thought to be determined by phosphatidylinositol-3-kinase (PI3K) to follow accumulation of phosphatidylinositol-3,4,5-trisphosphate (PIP3), resulting in the activation of Cdc42- and Rac-GEF. Localization of PIP3 can be monitored by the GFP-Akt-PH domain, which binds PIP3 and PIP2. In wild-type MEFs, PIP3 is highly concentrated at the leading edge, where accumulation of F-actin is induced. In contrast, actin filament organization at the leading edge does not occur in WAVE2-deficient MEFs, even though PIP3 levels are concentrated at the leading edge. It is also likely that WAVE2 localization at the leading edge is regulated by PIP3 (Oikawa et al 2004). WAVE2 binds specifically to PIP3 through basic regions in the centre of WAVE2, which is essential for localization of WAVE2 at lamellipodia.

These results indicate that although WAVE1 and WAVE2 are both involved in ruffle formation, they play different roles in membrane ruffle formation and cell migration.

References

Bear JE, Rawls JF, Saxe CL 1998 Scar, a WASP-related protein, isolated as a suppressor of receptor defects in late Dictyostelium development. J Cell Biol 142:1325–1335

Derry JM, Ochs HD, Francke U 1994 Isolation of a novel gene mutated in Wiskott–Aldrich syndrome Cell 78:635–644

Eden S, Rohatgi R, Podtelejnikov AV, Mann M, Kirschner MW 2002 Mechanism of regulation of WAVE1-induced actin nucleation by Rac1 and Nck. Nature 418:790–793

Innocenti M, Zucconi A, Disanza A et al 2004 Abi1 is essential for the formation and activation of a WAVE2 signalling complex. Nat Cell Biol 6:319–327

Lowenstein EJ, Daly RJ, Batzer AG et al 1992 The SH2 and SH3 domain-containing protein GRB2 links receptor tyrosine kinases to ras signaling Cell 70:431–442

Matuoka K, Shibata M, Yamakawa A, Takenawa T 1992 Cloning of Ash, a ubiquitous protein composed of SH2 and SH3 domains in human and rat. Proc Natl Acad Sci USA 89:9015–9019

Matuoka K, Shibasaki F, Shibata M, Takenawa T 1993 Ash/Grb-2, a SH2/SH3-containing protein, couples to signaling for mitogenesis and cytoskeletal reorganization by EGF and PDGF. EMBO J 12:3467–3473

Miki H, Takenawa T 1998 Direct binding of the verprolin-homology domain in N-WASP to actin is essential for cytoskeletal reorganization. Biochim Biophys Res Commun 243:73–78
Miki H, Miura K, Takenawa T 1996 A novel actin-depolymerizing protein N-WASP regulates the cortical cytoskeletal rearrangement in a PIP2 dependent manner at the downstream of tyrosine kinase. EMBO J 15:5326–5335
Miki H, Sasaki T, Takai Y, Takenawa T 1998a Induction of filopodium formation by a WASP related actin depolymeriziing protein N-WASP. Nature 391:93–97
Miki H, Suetsugu S, Takenawa T 1998b WAVE, a novel WASP-family protein involved in actin-polymerization induced by Rac. EMBO J 17:6932–6941
Miki H, Yamaguchi H, Suetsugu S, Takenawa T 2000 IRSp53 is an essential intermediate in the regulation of membrane ruffling by Rac and WAVE. Nature 408:732–735
Miura K, Miki H, Shimazaki K, Kawai N, Takenawa T 1996 Interaction of Ash/Grb2 via its SH3 domains with neuron-specific p150 and p65. Biochem J 316:639–645
Oikawa T, Yamaguchi H, Itoh T et al 2004 PtdIns(3,4,5)P3 binding is necessary for WAVE2-induced formation of lamellipodia. Nat Cell Biol 6:420–426
Rohatgi R, Ma L, Miki H et al 1999 The interaction between N-WASP and the Arp2/3 complex links Cdc42-dependent signals to actin assembly. Cell 97:221–231
Sakisaka T, Ito T, Miura K, Takenawa T 1997 Phosphatidylinositol 4,5-bisphosphate phosphatase regulates the rearrangement of actin filaments. Mol Cell Biol 17:3641–3649
Schlessinger J 2000 Cell signaling by receptor tyrosine kinases. Cell 103:211–225
Stradal TE, Rottner K, Disanza A, Confalonieri S, Innocenti M, Scita G 2004 Regulation of actin dynamics by WASP and WAVE family proteins. Trends Cell Biol 14:303–311
Suetsugu S, Miki H, Takenawa T 1999 Identification of two human WAVE/SCAR homologues as general actin regulatory molecules which associate with the Arp2/3 complex. Biochem Biophys Res Commun 260:296–302
Suetsugu S, Yamazaki D, Kurisu S, Takenawa T 2003 Differential roles of WAVE1 and WAVE2 in dorsal and peripheral ruffle formation for fibroblast cell migration. Dev Cell 5:595–609
Takenawa T, Miki H 2001 WASP and WAVE family proteins: key molecules for rapid rearrangement of cortical actin filaments and cell movement. J Cell Sci 114:1801–1809

DISCUSSION

Sheetz: Do the WAVE1-deficient cells spread on soft surfaces differently than they do on hard?

Takenawa: We only checked spreading on fibronectin-coated dishes. I don't know.

Sheetz: In our hands the ruffling often occurs when there is a low level of fibronectin, or when there isn't a rigid surface to pull on. Rigid surface pulling is important for establishing contacts.

Takenawa: We have no idea about that. But spreading speed is very important for the cells to adhere to a dish. We have tried to control the speed. Spreading speed of WAVE1-deficient fibroblasts is faster than that of wild-type cells. These cells hardly contact the fibronectin-coated dish at all.

Firtel: I have a question about the functional difference between WAVE1 and WAVE2. Is it a matter of localization of the two? They could perform an identical function. Or is it possible that the regulatory mechanisms that control the activity of WAVE1 and WAVE2 are different? It could be a combination of both. With WAVE2 we are talking about putting out outward protrusions and therefore the actin cytoskeleton is moving outward, while with WAVE1 it is movement towards the top of the cell and the actin cytoskeleton is arranged in a perpendicular direction. Is there a different mechanism of regulation, or is it a matter of localization of the two proteins? Can you do domain swaps between the proteins to address this question?

Takenawa: That is an important point. I think the basic function of WAVE1 and WAVE2 is the same, but the difference of localization determines what happens next. We don't know how localization is determined.

Firtel: Have you looked at the complex that is formed with WAVE1 and WAVE2? Are they the same proteins?

Takenawa: We think that the same proteins bind to both. We are now working on this. We used EGF-stimulated and EGF-non-stimulated lysates and precipitated them with anti-WAVE2 antibody. In this case, Arp2/3 complex was only precipitated in the presence of EGF. This is an important point, because these data show that Arp2/3 complex can bind to WAVE2 only when cells are stimulated by EGF.

Kaibuchi: I have a question concerning the localization of WAVE1 and WAVE2. Have you checked the interaction between Wave and microtubules?

Takenawa: We don't have any data on this. Haven't you done these experiments?

Kaibuchi: In collaboration with Dr Takenawa we are working on the role of WAVE1 and WAVE2 in axonal elongations. We have recently found that WAVE1 complexes with CRMP-2, which is a cargo receptor for kinesin. Have you looked at the movement of WAVE1 and WAVE2 along microtubules?

Takenawa: That is an interesting point. WAVE1 is translocated from intracellular vesicles to the plasma membrane area. In contrast, WAVE2 is constitutively present at the plasma membrane.

Borisy: I would like to try to get some clarity on the dorsal ruffles. As I understand them, dorsal ruffles are protrusions that fail to adhere to the substratum, which are then raised up and then move backwards because of retrograde flow. But in your pictures the dorsal ruffles show a kind of circular pattern. This makes me think that they are of a different nature than what have been referred to as ruffles in the past, coming from the periphery. How common are the dorsal ruffles, and do you have some suggestion for what they mean biologically? Are you sure they are on the dorsal surface?

Takenawa: When cells change their direction in 3D matrix, dorsal ruffles may be acting to effect this change in direction. Dorsal ruffles contain metalloproteases and secrete them to drive through.

Borisy: So your interpretation is that the dorsal ruffles are important for motility through 3D matrices.

Takenawa: Yes, I think so.

Borisy: How do you know that the ruffles are on the dorsal surface? Have you done scanning electron microscopy?

Takenawa: We haven't yet. We think that they are on the dorsal surface judging from the observation by light microscopy. But we do not have definite data to support this.

Borisy: I would suggest it is possible that they are on the ventral surface, and it is important that you check. I mention this because we and others have observed circular patterns such as those you have described in a variety of cultured cells. When we have looked at these by evanescent wave microscopy, they are on the ventral surface.

Nelson: If you use interference reflection microscopy, these circular patterns clearly form as transient focal adhesions with the substrate.

Borisy: Do you think what you have seen is the same as what Dr Takenawa is describing?

Nelson: They look very similar.

Takenawa: I think you are referring to the podosome. This structure is somehow similar to dorsal ruffles, though their location is different. They show the similar structure, consisting of circular actin filaments.

Alberts: Sara Courtneidge's lab has similar results in experiments using 3D reconstructions, looking at Src-induced invadopodia or podosomes. One of the problems here is that terms such as 'invadopodia' or 'podosomes' are used inappropriately in many cases. Experiments in Sara's lab have shown that they can invade into a matrix containing fibronectin. What is the functional physiological significance of these things?

Takenawa: Podosome forms at the substratum side. In contrast, dorsal ruffles form at the upper area of the cell. We already showed that podosome formation is regulated by N-WASP but not WAVE proteins. In contrast, dorsal ruffle formation is regulated by WAVE1. However, these structures have some similar characteristics.

D Lane: Thinking about invasion, have you tried to make tumours in these knockout mice? Or have you tried to make tumour cells which are deficient in WAVE1 or WAVE2? If so, do they difficulty in invading?

Takenawa: We have looked at this, and WAVE2 has an essential role in tumour cell invasion and metastasis, but not WAVE1.

Vallee: Mark McNiven and Pietro DeCamilli have shown evidence that at least one form of dynamin is associated with circular arc-like structures (Ochoa et al 2000, Krueger et al 2003). You are talking about WAVE1 being associated with punctate intracellular structures. Could an important difference between WAVE1 and WAVE2 be that WAVE1 is associated with membranous vesicles and undergoes a cycle of exposure to association with the surface and internalization that is not the case for WAVE2?

Takenawa: I think you are talking about the podosome, but not the dorsal ruffles.

Vallee: I am not sure that is entirely clear.

Takenawa: They are very similar but different in certain ways. Proteins such as dynamin, Rab5, Cotactin and Src are localized both in podosomes and dorsal ruffles.

Gundersen: You motioned that the punctate localization was membranous or vesicular. It has entered the discussion that one of WAVE1 or WAVE2 might be on vesicles. What is the evidence is that this punctate distribution really is vesicular, rather than a protein aggregate of some sort?

Takenawa: I only have evidence from immunofluorescence.

Nelson: Coming back to Rick Firtel's question, if WAVE1 and WAVE2 have similar functions but different localizations, have you tried overexpressing WAVE1 in your WAVE2$^{-/-}$ MEFs to see whether you can saturate the system?

Takenawa: Overexpression is problematic. On its own, it makes a cluster of actin filaments.

Harden: In terms of distinguishing between WAVE1 and WAVE2, you mentioned that you had preliminary results on the Abi/Nap1/Sra-1 complex. In the literature with regard to WAVE2, the indication is that this complex is making a positive contribution to WAVE2 function, whereas in one study done with WAVE1 the complex was inhibitory with regard to WAVE1 function (Eden et al 2002).

Takenawa: I think that complex formation is necessary for stabilizing WAVE2 and WAVE1 rather than activating or inhibiting them.

Harden: Have you done anything along these lines in terms of distinguishing between WAVE1 and WAVE2 function?

Takenawa: I have looked at this but I can't say anything here.

Balasubramanian: What is the phenotype in the mice that lack WAVE1 and WAVE2? Are they defective in gastrulation movements, for example? Do they have phenotypes only in migrating cells?

Takenawa: WAVE2 knockout is embryonic lethal due to a defect in angiogenesis (Yamazaki et al 2003). WAVE1 knockout mice showed defects in nervous system. There are two papers published on this (Sodering et al 2003, Dahl et al 2003). One reports that homozygous disruption of WAVE1 gene results in postnatal lethality; the other reports that WAVE1 knockout mice are born though their members are lower than expected. These mice showed sensorimotor retardation and reduced learning and memory.

Mostov: Speckled microscopy movies from Claire Waterman-Storer's group show that there are incredible movements of actin in different directions near the active leading edge. Are there particular differences in WAVE1- or WAVE2-deficient mice in this sort of process? Are specific processes interfered with by one or the other?

Takenawa: I think there are many proteins involved in cell movement and actin polymerization. These are just some of them.

Schejter: Given the similarities in structure and binding partners, can you suggest why these two homologues, WAVE1 and WAVE2, are differentially localized in the cell?

Takenawa: So far we haven't found different binding partners. We don't know.

Peter: You mentioned that in the WAVE2 knockouts, the actual polarization event is OK. But can polarity be maintained?

Takenawa: I don't know.

Sheetz: Are the dynamics of WAVE1 and WAVE2 in their various locations different? Have you done photobleaching experiments and looked at the recovery?

Takenawa: No.

Vallee: Are there defects in the rate of secretion or endocytosis in the WAVE1 knockouts?

Takenawa: I don't know.

Chang: Do you think that the primary function of these WAVE proteins is to regulate the Arp2/3 complex? Or are there other targets?

Takenawa: I think the Arp2/3 complex is the primary target, but these proteins are also working as an adaptor protein.

Chang: Do you think they could be affecting the Arp2/3 complex in slightly different ways?

Borisy: People have looked at whether the geometry of branching depends on the nature of the Arp activator. So far, they haven't found any differences in the branching. Probably not is the answer.

Takenawa: I think that basic functions of these proteins are the same: activation of Arp2/3 complex.

Borisy: Nevertheless, I think the point that Fred is raising is a good one. We do see cellular phenotypes as a result of knockout that affect relative adhesion. You interpreted the more rapid spreading as adhesion not being able to keep up with spreading. There is some difference in ability to adhere in the WAVE1 null cells. Also, you commented that there are more filipodia than normal in the WAVE2 knockouts. This is another phenotypic difference. Would you interpret this phenotype as arising through action on Arp, or is it possible that there is some other target that is also affected?

Takenawa: I don't know the precise mechanism how filopodia are formed instead of lamellipodia in WAVE2-deficient fibroblasts. Probably filopodia stand out in peripheral actin filaments, because lamellipodia are retreated in WAVE2-deficient fibroblasts.

Firtel: At some level there could be competing responses. If you suppress lamellipodium formation by knocking out WAVE2, it is possible that the mechanisms that are forming filipodia have a better way of forming.

Takenawa: I forgot to mention that PI3K inhibitors induce the formation of filopodia-like protrusions. PI3K inhibitors inhibit the translocation of WAVE2 to the plasma membrane. In this case, many filipodia are formed. We do not know whether this phenomenon is caused by the result of the inhibition of lamellipodia formation or disturbance of cortical actin filament formation. Massive actins may be used for assembling into filopodia by prevention of lamellipodia formation.

References

Dahl JP, Wang-Dunlop J, Gonzales C, Goad ME, Mark RJ, Kwak SP 2003 Characterization of the WAVE1 knock-out mouse: implications for CNS development. J Neurosci 23:3343–3352

Eden S, Rohatgi R, Podtelejnikov AV, Mann M, Kirschner MW 2002 Mechanism of regulation of WAVE1-induced actin nucleation by Rac1 and Nck. Nature 418:790–793

Krueger EW, Orth JD, Cao H, McNiven MA 2003 A dynamin-cortactin-Arp2/3 complex mediates actin reorganization in growth factor-stimulated cells. Mol Biol Cell 14:1085–1096

Ochoa GC, Slepnev VI, Neff L et al 2000 A functional link between dynamin and the actin cytoskeleton at podosomes. J Cell Biol 150:377–389

Soderling SH, Langeberg LK, Soderling JA et al 2003 Loss of WAVE-1 causes sensorimotor retardation and reduced learning and memory in mice. Proc Natl Acad Sci USA 100:1723–1728

Yamazaki D, Suetsugu S, Miki H et al 2003 WAVE2 is required for directed cell migration and cardiovascular development. Nature 424:452–456

A conserved role for myosin VII in adhesion

Margaret A. Titus

Department of Genetics, Cell Biology & Development, University of Minnesota, Minneapolis, MN 55455, USA

Abstract. The class VII myosins (M7) are expressed in a wide range of organisms. M7 mutants in mice, zebrafish and *Dictyostelium* exhibit phenotypes that reveal a role for M7 in adhesion in these highly divergent systems, suggesting a basic conservation of M7 function throughout evolution. M7s are characterized by the presence of two FERM domains in their C-terminal tail region, and deletion of either from the *Dictyostelium* M7 (DdM7) tail results in loss of function without affecting localization. A search for DdM7 binding partners has revealed that talin, an actin-binding protein that provides a key link between adhesion receptors and the actin cytoskeleton, interacts directly with DdM7. The phenotypes of the DdM7 and talin null mutants are highly similar, suggesting that these two proteins work co-operatively to maintain cell–cell and cell–surface contact and that this interaction may also be conserved throughout evolution.

2005 Signalling networks in cell shape and motility. Wiley, Chichester (Novartis Foundation Symposium 269) p 16–29

Cell migration and phagocytosis rely on controlled and specific adhesion to appropriate substrata. Binding of cell surface receptors to their ligand(s) triggers reorganization of the actin cytoskeleton through the local activation of small G proteins and kinases, and results in the formation of focal contacts or a phagocytic cup. These transient structures are sites where links between the actin cytoskeleton and receptors are organized, reinforced and ultimately disassembled. Several different actin binding proteins play a role in anchoring focal adhesions to the actin cytoskeleton; these include talin, vinculin and α-actinin. Talin is of particular interest as it binds directly to the cytoplasmic tails of integrins, the heterodimeric adhesion receptor, and regulates its activation (Nayal et al 2004). Talin also recruits focal adhesion kinase (FAK), vinculin and α-actinin to the growing adhesion complex and appears to have a key role in the generation of strong adhesion.

Myosins are ATP-dependent actin-based motor proteins that are usually considered to be actively moving cellular components or contracting actin bundles (Kieke & Titus 2003). For example, myosin V translocates pigment granules in melanocytes, and myosin II is responsible for constriction of the contractile ring during cytokinesis. One interesting development over the past few years has been the identification of a role for myosins in cell adhesion, either directly as part of receptor–cytoskeleton complexes or indirectly by (potentially) exerting tension on the adhesion complex. The first myosin implicated in adhesion is the class VII myosin (M7). Based on the analysis of M7 mutants from two different systems, mice and *Dictyostelium* (Maniak 2001), M7 was suggested to play a structural role instead of an expected transport role that might have been predicted based on the general properties of myosin superfamily members.

The M7s are characterized by a conserved myosin motor domain followed by 3–5 light chain binding IQ motifs, a region of predicted coiled coil (in most, but not all M7s) then a tandem repeat of a MyTH4/FERM domain (MyTH4, myosin talin homology 4; FERM, band 4.1, ezrin, radixin, moesin) that is often separated by an SH3 domain (Fig. 1A) (Kieke & Titus 2003). One of the two vertebrate isoforms, M7a, is found in the ear, eye, lung, kidney and testis where it is typically associated with specialized actin-based structures such as stereocilia (in the inner ear) or ectoplasmic specializations (in the testis) (Hasson et al 1995). This distribution is distinct from that of M7b, which is found only in epithelial cells of the kidney and intestine (Chen et al 2001). Baculovirus-expressed mouse M7a motor domain moves towards the barbed end of actin with an average velocity of 0.16 μm/s, a rate consistent with the measured rate of full-length immunoprecipitated M7a (0.19 μm/s) (Inoue & Ikebe 2002, Udovichenko et al 2002). The full-length expressed M7a is a dimer but it is of interest to note that the region of predicted coil is not sufficient for dimerization, suggesting that other elements in the tail are required. In spite of the fact the M7a is a dimer, the available kinetic and motility data do not clearly establish whether or not M7a is processive motor, but are consistent with it having a role in tension maintenance (Inoue & Ikebe 2002).

The indication that M7a plays a role in adhesion came first from the analysis of abnormalities in the cochlear hair cells in the mouse M7a mutant, *shaker 1* (Gibson et al 1995). Scanning electron micrographs revealed that the cochlear hair cells do extend stereocilia but instead of the individual stereocilia being linked together in the typical staircase bundle, they appear disarrayed (Self et al 1998). A similar disorganization of stereocilia is also found in the equivalent sensory cells of the zebrafish M7 mutant, *mariner* (Ernest et al 2000). The localization of M7a within the stereocilium near the region where stereociliary links are extended, structures referred to as ankle links, suggested that M7a may serve to anchor extracellular adhesion molecules to the actin cytoskeleton (Hasson et al 1997, Küssel-Anderman et al 2000). *Dictyostelium* mutants lacking M7 (DdM7) have defects in substrate adhesion — they exhibit reduced binding to particles at 4°C (a temperature at which engulfment does not occur), a decreased area of close

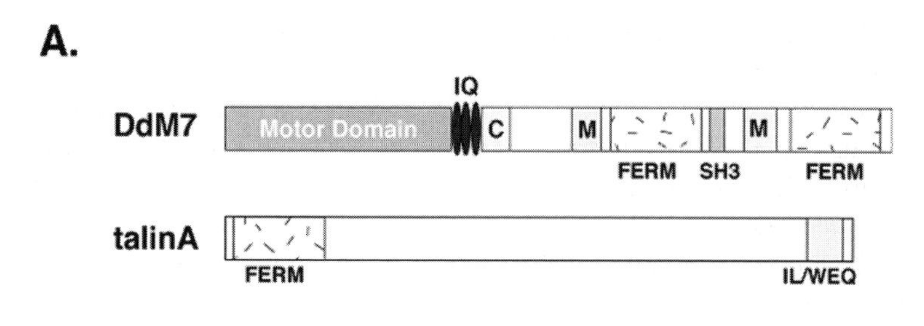

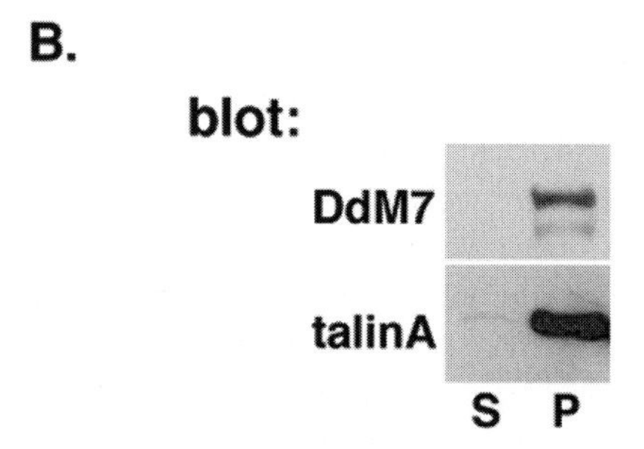

FIG. 1. TalinA is associated with DdM7. (A) Schematic diagrams of *Dictyostelium* M7 (DdM7, top) and talinA (bottom). The motor domain of DdM7 is shaded grey, the light chain-binding IQ motifs are dark ovals, the region of predicted coiled-coil is indicated by C, the MyTH4 domains by M, FERM domains by the stippled box and the SH3 domain by the lighter grey box and I/LWEQ indicates the position of the actin binding region of talinA. (B) Wild-type *Dictyostelium* expressing GFP tagged DdM7 were lysed with Triton X-100 and GFP-M7 immunoprecipitated (IP) using an anti-GFP antibody (Molecular Probes). The resulting IP supernatant (S) and pellet (S) was run on a 6% SDS gel and the samples transferred to PVDF membrane and the membranes probed for the DdM7 heavy chain or talinA. Note the presence of both DdM7 and talinA in the IP pellet.

contacts as visualized by interference reflection microscopy (IRM), and decreased calcium-sensitive cell–cell adhesion (Tuxworth et al 2001). The defect in adhesion results in reduced phagocytosis and aberrant motility. DdM7 is largely cytosolic with a small fraction present on the plasma membrane in regions of the cell actively extending an actin-filled projection (i.e. a pseudopod, a filopod or a macropinocytic crown) (Tuxworth et al 2001). DdM7 does not appear to act as a vesicle transport motor that brings receptor-containing vesicles to the plasma membrane, based on the presence of normal levels Phg1 and gp130 (molecules both implicated in adhesion and phagocytosis) on the plasma membrane. A key question emerging from studies in both mice and *Dictyostelium* is how M7 contributes to adhesion at the molecular level. Does this myosin bind to adhesion proteins directly or indirectly and is motor activity necessary for its function or is actin binding sufficient?

Elegant studies of both human and mouse M7a mutants has uncovered several M7a binding partners and led to a molecular model of the role of M7a in stereocilia organization. Patients with Usher syndrome type 1 (USH1) have combined deafness and blindness and they are classified into seven distinct subgroups, USH1A–G, based on symptoms (Petit 2001, Adato et al 2004). The systematic identification of the genes underlying these forms of USH1 and analysis of the corresponding mouse mutants has led to the discovery of a functionally related set of five proteins that serve to link adjacent stereocilia by anchoring transmembrane proteins through the plasma membrane to the action core. *USH1B* encodes M7a, and it interacts with the protein encoded by *USH1C*, a PDZ-containing protein named harmonin. Harmonin, in turn, interacts both with the protein encoded by *USHIG*, another scaffolding protein named Sans, as well as cadherin 23 (*USH1D*) and protocadherin 15 (*USH1F*). M7a also binds to Sans as well as a novel linker protein, vezatin that interacts with α-catenin that, in turn, binds to cadherins at cell–cell junctions (Küssel-Anderman et al 2000). The data indicate that M7a is linked to calcium-dependent adhesion molecules, i.e. the cadherins, via interactions with distinct scaffolding proteins. It is not known if these proteins are present all together in one large supramolecular adhesion complex or if there is differential association of various components depending on where the proteins are located along the length of the stereocilia.

Progress in identifying the connections between M7a and other USH proteins in the specialized cells of the ear prompts the question of whether or not these interactions are unique to the ear or can be generalized to other cell types where M7a (or M7b) is found or even to other organisms that express M7s, such as flies, worms and amoebae. A search for M7 binding proteins in *Dictyostelium* has been undertaken to address this issue. A GFP-tagged DdM7 is competent to rescue the null mutant phenotype and this was exploited in an initial search for binding partners. GFP-DdM7 was immunoprecipitated (IP) from either physically or detergent lysed cells using a commercially available antibody (Fig. 1A; R. I. Tuxworth, S. Stephens, Z. C. Ryan, M. A. Titus, unpublished results). The resulting IP pellet contains two distinct bands, one of ~300 kDa and another that is ~250 kDa. The upper band is GFP-DdM7 and the lower band a *Dictyostelium* homologue of talin, talinA.

Talins are large, homodimeric proteins characterized by the presence of an N-terminal FERM domain and a C-terminal actin binding region (I/LWEQ) (Fig. 1A) (Nayal et al 2004). The talin FERM domain binds both the cytoplasmic tails of integral membrane proteins such as integrins as well as phospholipids, specifically phosphatidylinositol-4,5-bisphosphate (PIP2). Talin is an integral component of the focal adhesion complex in mammalian cells where it binds

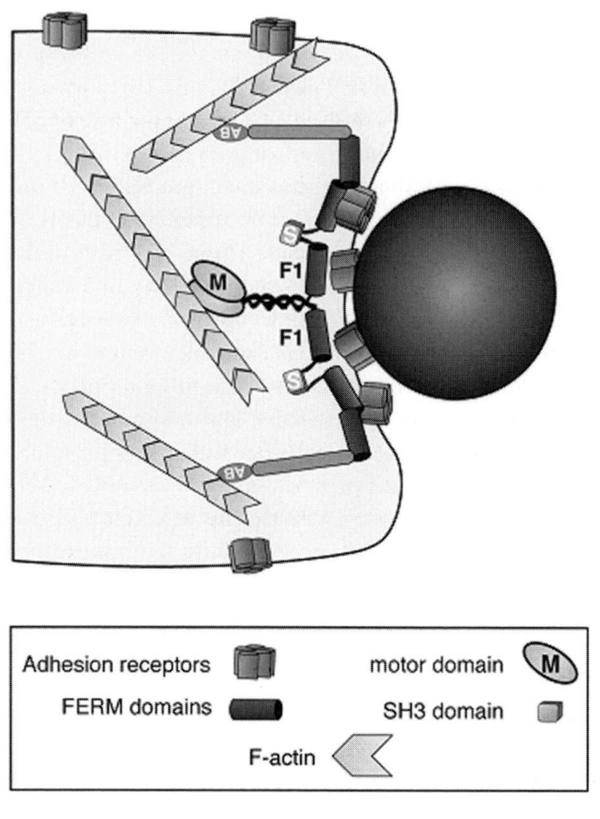

FIG. 2. Model illustrating the speculated roles of *Dictyostelium* M7 and talinA in linking adhesion receptors to the actin cytoskeleton. Shown is a region of a *Dictyostelium* cell binding a particle. DdM7 and talin serve to organize a potential high avidity adhesion receptor/talinA/DdM7 complex through interaction of the FERM domains with adhesion receptors and anchoring them to the actin cytoskeleton via the DdM7 motor domain and the I/LWEQ actin binding region of talinA.

FAK and has been shown to recruit PIPKIγ661, a PIP kinase, to focal adhesions in addition to associating with other actin binding proteins (Nayal et al 2004). Talin mutants from a range of organisms are defective in adhesion. *Dictyostelium* talinA appears to be largely cytosolic although a portion is localized to the plasma membrane where it is concentrated at the leading edge of migrating cells as well as in filopodia (Kreitmeier et al 1995). The talinA null mutant phenotype is similar, but not identical to that of the DdM7 null mutant. Loss of talinA results in defects in the uptake of bacteria, a substantial reduction in regions of close contact with the substrate and a loss of calcium-dependent cell–cell adhesion, phenotypes that are quite similar to those observed in the DdM7 null mutant (Niewöhner et al 1997).

Analysis of mutant cells expressing GFP-tagged DdM7 or talinA reveals that talinA is not necessary for the localization of DdM7 or vice versa. Thus, talinA is not a receptor for DdM7 (and DdM7 is not a receptor for talinA), suggesting that each binds to its own docking protein on the plasma membrane (Tuxworth et al 2005).

The finding that DdM7 interacts with talinA is of interest in light of a report indicating that talinA is nearly absent in a DdM7 null mutant (Gebbie et al 2004). This finding suggests that DdM7 is necessary to stabilize talinA directly and one could then infer that the reported DdM7 null phenotype is actually due to the loss of talinA. However, it should be noted that the phenotypes of the DdM7 and talinA null mutants are not identical. The DdM7 null fails to extend filopodia while the talinA null mutant still makes filopodia (Tuxworth et al 2001). Furthermore, the talinA mutant has a mild cytokinetic defect while the DdM7 null mutant does not (Niewöhner et al 1997, Hibi et al 2003, Gebbie et al 2004). Also, a more recent analysis of the talinA levels in the originally described DdM7 null mutant reveals that talinA is still present, although there is a decrease in overall levels to 30–50% of that found in control cells (S. Stephens, M. A. Titus, unpublished data). Whether DdM7 protects talinA from degradation or controls talinA levels indirectly (i.e. via feedback control of transcription or translation) is not yet known. However, taken together, these data suggest that the DdM7 and talinA are functionally linked in some respects but that each has unique roles.

The contribution of the individual M7 tail domains to its function is poorly understood. Although missense mutations in the M7 tail have been identified in human Usher syndrome type 1B patients, *shaker 1* mice and *mariner* zebrafish (Liu et al 1998, Ernest et al 2000), it is not known if these changes result in destablization of the protein, reduced interaction with binding partners and/or mislocalization. A tail overexpression approach has been used to explore the role of the FERM domains in M7 localization and function as the FERM domains of ERM family members and talin have been shown to interact with integral membrane proteins. The DdM7 tail is localized similar to the full-length DdM7, it is largely cytosolic with a portion found on the plasma membrane at actively extending regions of the cell (R. I. Tuxworth, S. Stephens, M. A. Titus, unpublished data). Expression of the tail in wild-type cells results in reduced phagocytic activity and preliminary results reveal that deletion of either the N-terminal or the C-terminal FERM domain eliminates this dominant negative phenotype while not altering the localization of the DdM7 tail to the plasma membrane (Titus et al 2004). These results suggest that both FERMs must be able to interact with the DdM7 binding partner, presumably on the plasma membrane, in order to cause the dominant negative phenotype. Experiments to define how tail overexpression results in reduced phagocytosis as well as the roles of other M7 tail domains to localization/function of the tail are ongoing.

A comparison of the gross phenotypes of mice and *Dictyostelium* M7 mutants suggests that M7 has a conserved role in linking integral membrane adhesion receptors to the underlying actin cytoskeleton. This is quite remarkable given such a large evolutionary distance between vertebrates and *Dictyostelium*. Is there really strong conservation of M7 function throughout evolution? Some aspects of the mouse *shaker 1* phenotype support a model in which M7 transports components of the adhesion complex to their correct cellular location. At least one *shaker 1* allele retains some stereocilia organization yet the mutant mice are profoundly deaf (Kros et al 2002). One of these mutant alleles ($Myo7a^{6J}$) exhibits defects in the gating of transduction channels due to an alteration in their resting tension (Kros et al 2002). Two models have been proposed to account for this observation: M7a may provide the force necessary for maintaining resting tension of the channel but an alternative hypothesis is that M7a has a role in translocating other proteins that have a role in transduction up to the tip of the stereocilia where the channels are located (Boëda et al 2002, Kros et al 2002). Consistent with a transport model for M7, several of the known human and mouse M7a missense mutations reside in the motor domain (Liu et al 1998) and it has been shown recently that harmonin is not correctly localized to the tips of stereocilia in *shaker 1* mice (Boëda et al 2002). Transport and structural roles for M7 are not mutually exclusive as there may be specific regulatory mechanisms, such as interactions with different subsets of binding partners or changes in tension applied to the motor, that could serve to switch M7a from one mode to another. An example of this type of switch has recently been shown for M6. Application of tension to a dimeric M6 in an optical trap results in a conversion from a motor protein to a tether or anchor (Altman et al 2004) and one can envision a similar scenario for M7.

An investigation of M7 function in another model system, worms, has been undertaken to address the conservation of M7 function throughout evolution and provide another system in which to explore the potential motor versus structural role of M7. The *Caenorhabditis elegans* genome contains a single M7 gene, *hum-6* (heavy chain of an unconventional myosin) (Baker & Titus 1997). HUM-6 is quite similar to other M7 proteins, except that it lacks the region of predicted coil at the beginning of the tail and SH3 domain that separates the two MyTH4/FERM domains. Injection of a transgene carrying a transcriptional fusion of the *hum-6* promoter to *gfp* results in broad, but not ubiquitous expression of GFP in the developing worm embryo. GFP is found in the pharynx and around the periphery, where the developing muscle and hypodermis reside, of the comma stage embryo (J. P. Baker, M. A. Titus, unpublished data). A mutant allele that lacks the 5′ coding region of the *hum-6* gene has been characterized (J. P. Baker, G. Moulder, R. Barstead, M. A. Titus, unpublished data) and, surprisingly, develops normally. This result is at odds with what has been found for M7 mutations in all other organisms where either loss of M7 expression or mutations in the motor domain result in notable phenotypes. Analysis of the site of the deletion and mRNA expressed in the mutant worm reveals the expression of a truncated mRNA that encodes a portion of the motor domain and the entire tail region is produced in the mutant (J. P. Baker, M. A. Titus, unpublished data). This would lead to the production of a near full-length HUM-6 that lacks motor activity, a finding that indicates that the M7 tail alone is sufficient for function. The *C. elegans* talin is UNC-35 (Cram et al 2003) and one intriguing possibility, suggested from the finding that M7 and talin interact in *Dictyostelium*, may be that UNC-35 compensates for the loss of a potential link between adhesion receptors and the actin cytoskeleton if the two proteins are in a complex together or acting in similar sites of cell–cell contact.

Analysis of M7 in two highly divergent organisms has revealed a potentially conserved role for this motor protein in adhesion. It also establishes that M7 interacts with proteins that link the cytoplasmic tails of adhesion receptors to the underlying actin cytoskeleton. Several aspects of M7 function remain to be determined. Among these, it will be of interest to define the requirement for the motor domain in either active transport and/or actin binding. Another important question is how the interaction of M7 with its binding partners is regulated. The localization of DdM7 is quite dynamic, suggesting the existence of proteins that must spatially and temporally control where M7 functions. Ongoing studies in both *Dictyostelium* and *C. elegans* should provide additional insight into both of these outstanding questions.

Acknowledgements

I would like to thank all of the Titus Lab members who have contributed to this work over the years, especially Drs Richard Tuxworth and Jeff Baker, Gregory Addicks, Zachary Ryan and Stephen Stephens. Thanks also to Drs Ann Rougvie and Lihsia Chen for helpful advice during the course of our *C. elegans* work, Gary Moulder and Dr Bob Barstead (OMRF) for providing the *C. elegans* M7 mutant and Günther Gerisch (MPI, Martinsried) for generously providing the talinA antibody. Work in the Titus laboratory is supported by a grant from the NIH.

References

Adato A, Michel V, Kikkawa Y et al 2004 Interactions in the network of Usher syndrome type 1 proteins. Hum Mol Genet 14:347–356

Altman D, Sweeney HL, Spudich JA 2004 The mechanism of myosin VI translocation and its load-induced anchoring. Cell 116:737–749

Baker JP, Titus MA 1997 A family of unconventional myosins from the nematode *Caenorhabditis elegans*. J Mol Biol 272:523–535

Boëda B, El-Amraoui A, Bahloul A et al 2002 Myosin VIIa, harmonin and cadherin 23, three Usher 1 gene products that cooperate to shape the sensory hair bundle. EMBO J 21:6689–6699

Chen ZY, Hasson T, Zhang DS et al 2001 Myosin VIIb, a novel unconventional myosin, is a constituent of microvilli in transporting epithelia. Genomics 72:285–296

Cram EJ, Clark SG, Schwarzbauer JE 2003 Talin loss-of-function uncovers roles in cell contractility and migration in C. elegans. J Cell Sci 116:3871–3878

Ernest S, Rauch GJ, Haffter P et al 2000 *Mariner* is defective in *myosin VIIA*: a zebrafish model for human hereditary deafness. Hum Mol Genet 9:2189–2196
Gebbie L, Benghezal M, Cornillon S et al 2004 Phg2, a kinase involved in adhesion and focal site modeling in *Dictyostelium*. Mol Biol Cell 15:3915–3925
Gibson F, Walsh J, Mburu P et al 1995 A type VII myosin encoded by the mouse deafness gene *shaker-1*. Nature 374:62–64
Hasson T, Heintzelman MB, Santos-Sacchi J, Corey DP, Mooseker MS 1995 Expression in the cochlea and retina of myosin VIIa, the gene product defective in Usher syndrome type 1B. Proc Natl Acad Sci USA 92:9815–9819
Hasson T, Gillespie PG, Garcia JA et al 1997 Unconventional myosins in inner-ear sensory epithelia. J Cell Biol 137:1287–1307
Hibi M, Nagasaki A, Takahashi M, Yamagishi A, Uyeda TQP 2003 *Dictyostelium discoideum* talin A is crucial for myosin-II independent and adhesion-dependent cytokinesis. J Musc Res Cell Motil 25:127–140
Inoue A, Ikebe M 2002 Characterization of the motor activity of mammalian myosin VIIA. J Biol Chem 278:5478–5487
Kieke MC,Titus MA 2003 The myosin superfamily—an overview. In: Schliwa M (ed) Molecular motors. Wiley VCH, Weinheim, p 3–44
Kreitmeier M, Gerisch G, Heizer C, Müller-Taubenberger A 1995 A talin homologue of *Dictyostelium* rapidly assembles at the leading edge of cells in response to chemoattractant. J Cell Biol 129:179–188
Kros CJ, Marcotti W, van Netten SM et al 2002 Reduced climbing and increased slipping adaptation in cochlear hair cells of mice with *Myo7a* mutations. Nat Neurosci 5:41–47
Küssel-Anderman P, El-Amraoui A, Safieddine S et al 2000 Vezatin, a novel transmembrane protein, bridges myosin VIIA to the cadherins-catenins complex. EMBO J 19:6020–6029
Liu XZ, Hope C, Walsh J et al 1998 Mutations in the myosin VIIA gene cause a wide phenotypic spectrum, including atypical Usher Syndrome. Am J Hum Genet 63:909–912
Maniak M 2001 Ushering in a new understanding of myosin VII. Curr Biol 11:R315–R317
Nayal A, Webb DJ, Horwitz AF 2004 Talin: an emerging focal point of adhesion dynamics. Curr Opin Cell Biol 16:94–98
Niewöhner J, Weber I, Maniak M, Müller-Taubenberger A, Gerisch G 1997 Talin-null cells of *Dictyostelium* are strongly defective in adhesion to particle and substrate surfaces and slightly impaired in cytokinesis. J Cell Biol 138:349–361
Petit C 2001 Usher syndrome: from genetics to pathogenesis. Ann Rev Hum Genet 2:271–297
Self T, Mahony M, Fleming J et al 1998 Shaker-1 mutations reveal roles for myosin VIIA in both development and function of cochlear hair cells. Development 125:557–566
Titus MA, Stephens S, Ryan ZC 2004 Mapping functional domains in the M7 tail region. Mol Biol Cell 15S:222a
Tuxworth RI, Weber I, Wessels D et al 2001 A role for myosin VII in dynamic cell adhesion. Curr Biol 11:318–329
Tuxworth RI, Stephens S, Ryan ZC, Titus MA 2005 Identification of a myosin VII-talin complex. J Biol Chem in press
Udovichenko IP, Gibbs D, Williams DS 2002 Actin-based motor properties of myosin VIIa. J Cell Sci 115:445–450

DISCUSSION

Drubin: Have you have made the double mutant in *Dictyostelium*? Do you see any additional phenotype? Also, have you made a myosin/ATPase mutant in either *Dictyostelium* or in worms?

Titus: We have made a *Dictyostelium* M7/talinA mutant. It looks as you would predict. We measure the phenotype initially by phagocytosis, which is used as a surrogate for an adhesion assay, because if beads don't bind they are not internalized. There is a distinct difference between the M7 and talinA mutants. The double mutant looks the same as the talinA mutant by our surrogate adhesion assay. The double mutants don't make filopodia. We have not made an actin-binding site mutation in either the *Dictyostelium* or *C. elegans* M7. The *C. elegans* experiment is difficult because the gene is quite large. In our rescue attempts we only see dead embryos. We think this is because when we inject the cosmid into worms they make large extrachromosomal arrays, and worms are very sensitive to dosage. Now we are trying to make the *C. elegans* motor domain deletion in the *Dictyostelium* myosin.

Luca: Do you get an overexpression for M7 and talin? Does overexpression of one affect binding of the other? Is there a specificity for whatever membrane protein you are binding?

Titus: We have only done overexpression of M7. If we overexpress the FERM domains they are not localized and we don't see any defects in apparent motility, adhesion or phagocytosis. If we overexpress the tail alone we get an inhibition of phagocytosis. We have been trying to take out and add back domains to figure out the minimal unit needed for this to occur. We haven't got there yet. We started with the FERM domains, and we have worked our way through the MyTH4 domains. One of the regions we are now focusing on is the coiled coil.

Borisy: Could you remind us of the relationships between M7 and M10? Does *Dictyostelium* just have M7?

Titus: Dictyostelium only expresses M7. M10 is only found in higher eukaryotes.

Borisy: When you are studying functions of M7 in *Dictyostelium* and *C. elegans*, and speculating on conserved functions, are you thinking that it is serving the same kind of role as M10 does in mammals?

Titus: It probably serves the same core functions as M10. M10 is ubiquitously expressed in higher eukaryotes, whereas M7 has a highly restricted expression pattern. M7 has a lot of the basic functions that M10 has.

Nelson: I understood that when you look in a M7 null mutant background talin can be localized normally, and vice versa. So what is the complex between talin and M7?

Titus: I don't know. This has been challenging and frustrating, because we have done all the obvious things. We are now interested in whether or not the conformation of the two proteins changes when they bind to the membrane and bind to the receptor proteins.

Nelson: Is the complex you isolated from the cytosol or membranes?

Titus: Both. Stephen also looked at cells making filopodia, as well as cells which were moving chemotactically and were therefore highly polarized to see whether there was some change in the complex. The immunoprecipitation (IP) looks identical and there are no additional proteins visible by silver staining. We see the same thing time and time again. I think we are looking at some kind of mechanical link between these proteins which might change on the membrane.

Alberts: The IP you are seeing is fairly substantial. Is this inducible? Is there a situation where you can induce the assembly of the complex?

Titus: We hoped that we might come up with something like that. We put the *Dictyostelium* through their paces in different conditions and we always see the same thing. It could be that there is something about our IP conditions that is not permitting other binding partners to come along for the ride. One suspected binding partner was VASP, because Rick Firtel's lab has nicely shown that the VASP phenotype is also similar to the M7 phenotype. We looked for an interaction between VASP and M7 but didn't find any. It is mysterious.

Cai: Is the GFP-M7 staining in the steady-state situation? Are there changes over time?

Titus: It appears to be pretty dynamic. If the cell encounters a particle, for example, the M7 will stay at the plasma membrane as the particle is being engulfed. Before this is completed the M7 comes off.

Cai: Does this remain the same in the mutant?

Titus: We didn't look at the dynamics of M7 in the talinA mutant. We looked at total membrane preps, and showed that the level was unchanged. One thought was that we were getting too much, or not enough on the membrane quantitatively, but as far as we can tell at the biochemical level it is unchanged.

Cai: The other thing is that the *hum-6* mutant may not be very good. Have you used RNAi to see whether there is an effect?

Titus: Yes, and there is no RNAi phenotype. What you are bringing up is a good point. We now have a second mutant, *ok632.* The deletion introduces a premature stop. We find we can never get adult homozygous deletion worms. If we look at our plates, we see the worms arrested at what is called the twofold stage. This is where muscle and hypodermis attachment must occur for the worm to begin to elongate. These worms still twitch but the movement is rather irregular, so there is some anchoring of the muscle to the hypodermis. This is probably a null mutant and it gives genetic evidence that the other mutant is not a null. We have outcrossed these worms and we know that there is not a background modifier or mutation.

Sheetz: In mouse fibroblasts we have been looking at talin1 function. It appears to have a role in helping organize scaffold proteins for subsequent functions. One of the possibilities is that the mechanical organization caused by binding to a surface then confers to the talin a different configuration, which then allows M7 to be activated in that region. This gives rise to a mechanical signal that would flatten the membrane onto the surface as is needed in these systems. Is that a viable alternative hypothesis?

Titus: That is perfectly consistent with what we are seeing. Perhaps in order for this process to be more efficient there is a complex of the two proteins, ready to go in the cytosol. This gets onto the membrane and talin binds scaffolding proteins. This is consistent with our notion that talin is closer to the membrane and makes a bigger contribution to adhesion than M7. After talin binds its receptor, M7 then changes its conformation. This is what we think is going on.

Harden: Is there any genetic or biochemical evidence for an interaction of integrin with M7 in worms?

Titus: Not at present.

Vallee: I am interested in the complex. If you IP in both directions is there an excess of free talin over myosin? This is relevant to the whole issue of the different phenotypes. It is hard to imagine that there is a stoichiometric complex between the two, and that is all you are getting: this complex.

Titus: If we IP, we analyse both the supernatants and pellets. The partner is not present in the supernatants. When we do our IP, everything is coming down. In order to do these IPs we are probably mildly overexpressing the GFP-tagged protein (either M7 or talinA). This could be sopping up everything. There could be some free in wild-type cells. However, we tried to look at this by sucrose gradient centrifugation, and if we fractionate the cytosol the complex sediments at about 14S in wild-type cells. In the two mutants, both proteins shift up into the very lightest fractions. It looks to us like both proteins are fully complexed *in vivo*. When you take one out it goes into a lighter fraction. My guess is that they are stoichiometrically associated with each other.

Firtel: David's question was trying to address the motor function. You have a deletion in motor function in the worm. In your model, what is the purpose of the motor function? Is it for regulatory release of the complex from the actin cytoskeleton? Why do this with a motor domain, if we are not going to be moving along the actin?

Titus: The results with our worm motor domain mutant say that you don't need to motor to be a myosin. You can use the motor as a fine-tuning tension-providing element in a complex of proteins. You might have links to actin that you want to tighten, or you might want to respond to external stress, and the motor would allow you to do this quickly. The kinetics of this motor haven't been fully characterized. The initial evidence is that is a short duty motor that might be suitable for the maintenance of tension.

Borisy: There are other examples in the literature where myosin function can be inherited without an intact motor domain. In cytokinesis there are examples where headless mutations can still support trafficking of filaments to the equatorial plane.

It still localizes and supports that trafficking. This is another example where a motor function isn't required, yet it is a bona fide myosin.

Takenawa: Have you looked at binding to the FERM domain?

Titus: We haven't done this. The FERM domain of myosin is missing canonical sequences for phosphoinositide binding that are present in Talin.

Takenawa: Mammalian cells have phosphatidylinositol-3,4,5-trisphosphate (PIP3), which plays an important role in protein translocation. Does *C. elegans* have this system?

Titus: I didn't look for it in worms. I looked for PIPKIγ661, a PIP kinase that is recruited by talin in mammalian cells, in *Dictyostelium* and could not find the homologous kinase.

Borisy: I'd like to ask about the interpretation of your phenotype at the cellular level. You say you have a deficiency in attachment of the hypodermis to the muscle. What is the origin of that deficiency? Another point you made is that at the cellular level there is a deficiency in filopodial extension. Do you connect these two? Is it possible that the deficiency of tissue attachment is related to inability to make filopodia? You said that the M7 talin is localized at the tips of filopodia as well as at the perimeter of the cell. Can you speculate on the connection between filopodia extension and the tissue phenotype?

Titus: In *C. elegans*, when the muscle develops and starts to interact with neurons, there are structures called muscle arms which are a little bit like filopodia that come out and make direct contact with the neurons. I don't think they are fine enough to be considered to be true filopodia, but they probably are some kind of actin-driven protrusion that has to go out and seek target neurons. One possibility is that they are not getting out efficiently and finding their binding partners. This is highly speculative. We would really like to look at development of muscle arms *in vivo*.

Borisy: What is the role for M7 in filopodial extension?

Titus: I don't know. At first we thought that it might be localizing VASP to sites where filopodia come out, but I don't think this is correct. The other model is that M7 is organizing the actin filaments at the base and providing some kind of tension.

Chang: M7 could be associating with another myosin. It might be functionally redundant in that manner.

Titus: On the basis of localization and IP data we have no evidence for this.

Chang: Has anyone seen M7 GFP particles moving on actin filaments?

Titus: It is hard for us to watch the filopodia grow out, because the filopodia are coming out in three dimensions. These are not big flat cells where we can watch them shoot out more easily. We have the impression that GFP M7 is coming out with the filopodium, but it is hard to track because they are zipping in and out of the plane of focus. We haven't tried to flatten the cells.

Borisy: Fred Chang is presumably asking this question because Richard Cheney's lab has observed M10 spots moving along actin filaments. Even if you haven't observed M7 doing the same, if you say that it is moving with the filopodial tip as it elongates, given that the elongation process occurs by the addition of actin subunits at the tip, you would still conclude that the M7 spot has to move along the filaments to remain with the tip. It has to adjust its position because new subunits are being interpolated at the tip of the filopodium.

Titus: Unless it is tethered to the plasma membrane. We tried to look for interaction with other proteins that are involved in actin polymerization. For filopodia this doesn't make sense, but we thought p116 might be interacting with the SH3 domain of M7. We found no evidence for this.

Luca: Do you need actin filaments to maintain cortical localization of M7?

Titus: I am pretty sure that they are independent.

Borisy: You said that a VASP knockout gives a similar phenotype. Do you infer that VASP is another molecule needed for filopodium formation?

Titus: Mitsuo Ikebe's lab has just published a report showing that VASP binds to the tail of M10 (Tokuo & Ikebe 2004). That is why we went back to look to see whether M7 interacted with it. It didn't.

Borisy: What is the phenotype of a fascin knockout in worm?

Titus: There isn't a fascin homologue in worms as far as I know.

Borisy: This would be another way of preventing filopodium formation.

Alberts: Most of these mutations that you showed are loss-of-function. Are any of them dominant in any way? How much of this is due to a scaffolding function?

Titus: The mutant that is dominant is the tail. If you overexpress the tail in wild-type this acts as a dominant negative in *Dictyostelium* and these cells don't appear to make filopodia. All the data we have suggest it is scaffolding and not motoring. The tail is organizing something in the membrane, and it may be doing it by its interaction with talin.

Sheetz: What is the evidence that the motor is not involved?

Titus: Our worm mutant.

Reference

Tokuo H, Ikebe M 2004 Myosin X transports Mena/VASP to the tip of filopodia. Biochem Biophys Res Commun 319:214–220

General discussion I

Firtel: A number of people have been talking about circular waves. I'm curious about how circular waves are formed. And how are they regulated in relationship to this undefined upstream signalling pathway? How do you get a circular wave of actin polymerization?

Alberts: There is evidence to indicate that it could be an evanescent wave. In polarized cells, it would focus at the leading edge as part of a feed-forward mechanism that is dependent upon the focused generation of specific lipids that drive the localization of key molecules. In some cases polarization would be dependent upon this machinery to focus the leading edge. Movies using green-fluorescent protein (GFP)-tagged pleckstrin homology domains recruited to waves or leading edges are quite compelling. For the waves, it looks almost as if someone has dropped a pebble on a pond.

Nelson: In epithelial cells, which are cells that are contacting each other, there may be relatively few ventral ruffles. Then one cell will start off forming a wave and then the next cell will start forming one as well. This suggests that there is an activity that is signalling between the cells.

Mostov: Is that signalling by tension, or cell surface receptors, or gap junctions?

Nelson: I would guess that it is by gap junctions.

Firtel: Basically, the bottom line is that we don't know how they are initiated.

Borisy: We can't agree on whether they are on the top or bottom of the cell.

D Lane: Someone earlier raised the issue of the relationship with podosomes. I'd like to hear more about that. This is the work by Sara Courtenidge (Abram et al 2003).

Alberts: The argument has been about the nature of podosomes and where they occur. Her recent experiments have been done using fluorescently labelled matrix. It is possible to take the cell, fix it and reconstitute the image. She can see podosomes digging into the membrane, and at the base of those dips are matrix metalloproteases (MMPs) which are digesting the membrane. This is one good example of where podosomes can be seen experimentally. But then again, when you see these dorsal ruffles or pond ripples occurring on cells, they look very different from what are considered bona fide podosomes.

Sheetz: This discussion reminds me that many times we are talking about different structures under the same label, lumping a lot of different things together. We should spend some time defining what we mean by various statements. A wave of lamella formation on the upper surface can often be likened to the coronin story in *Dictyostelium*. This podocyte formation is going to be something quite different, I feel.

Borisy: That is an important point. How far back should we go? In listening to the discussion about localization of markers at the leading edge, I was wondering whether we have a common definition of the leading edge. We have all seen localization of markers at the cell perimeter, but this is not the same as the leading edge. We use the term 'lamellipodia': do we have a common definition for this? Or even for 'filopodia'? Are we lumping a number of different structures together under the same name? Abercrombie started this field about 50 years ago, and he was the one who introduced the terms 'lamella' and 'lamellipodia' for specific structures involved in cell crawling. He referred to the lamellum as a sheet-like structure in advance of the nucleus in the direction that the cell was migrating. Most of the flat part of the cell comprises the lamellum. Beyond the lamellum was the lamellipodium which, in contrast, referred to a dynamic, rapid exploratory structure, just a few microns deep at most. The lamellipodium underwent cycles of protrusion and withdrawal. It would sometimes go forward and sometimes retreat. Abercrombie likened the advance of the fibroblast to someone walking three steps forwards and two steps back. The lamellipodium cycled between states of protrusion and withdrawal. When we look at a static image I wonder whether the localization of a particular molecule at the perimeter depends on whether the leading edge is in a state of advance or a state of withdrawal. Another point is that some investigators deliberately create conditions to quiet cells. For example, a common experimental paradigm is to serum-starve cells, rendering them immotile but responsive to growth factors. These quiescent edges are in a very different state from the perimeter of a cell in protrusion or withdrawal. Then we have the behaviour of the cell edge or cell surface when it is endocytosing or when it is secreting. This may be related to what has been referred to as podosomes or invadopodia, or dorsal and ventral ruffles. These are just a number of ways in which we refer to the surface behaviour and edge behaviour of cells. With regard to filopodia, these were originally defined as needle-like processes based on their morphology, but there are actually many kinds of slender cellular processes. Are they all the same? Microvilli of epithelial cells lining the intestinal brush border and microvilli on cultured cells go by the same name but actually have different internal organization. Microspikes, ribs and filopodia are terms that have sometimes been used interchangeably. Within the class of filopodia, there are different categories. Some of the filopodia of the neural growth cones are branched, some are parallel, some are said to contain microtubules that transport small transmitter vesicles while others do not. These are all slender processes. Sometimes they are formed by extension, other times they are formed by cell membrane retraction. Are they equivalent? We need more and better molecular information before a clear answer can be given.

Nelson: Another distinction worth making is whether we are talking about cell movement in two dimensions or three.

Borisy: Coming back to Dr Takenawa's presentation, he talked in part about a phenotype of motility in a 3D matrix. The WAVE2-deficient cells are also deficient in motility through a 3D matrix. I raise this because WAVE2 has been shown to be important for this perimeter behaviour. Yet this kind of cell is also deficient in migrating in three dimensions. The labs of Friedel and Condeelis have looked at motility of cells in a 3D matrix, and refer to this as being 'amoeboid-like'. As I understand amoeboid motility, it refers to mass transport of cytoplasm within the cell as the cell locomotes forwards. The general point I want to make is that some investigators believe that cells express a different motility mechanism for going through a 3D matrix.

Sheetz: This flies in the face of what Fred Grinnell has been seeing in terms of dendritic arborizations of fibroblasts in 3D collagen matrices.

Nelson: My understanding of Peter Friedl's experiments (Wolf et al 2003) is that when he removes all the metalloproteinase activity in cells he gets a transition from one growth form to another.

Mostov: I agree. Amoeboid is one end of the spectrum; the other is the dendritic. You can move on that spectrum by blocking things such as MMPs.

Alberts: Chris Marshall has similar experiments, where he can get cells to switch by using Rho kinase inhibitors.

Nelson: Friedl made the argument that some MMPs are secreted in a polarized manner at the leading edge of cells, and this is what causes breakdown of the 3D matrix and allows cells to migrate through the matrix (see Wolf et al 2003).

Borisy: This is a fundamental question: do cells have another motility mechanism? Do they have an amoeboid mechanism for going through a 3D matrix and a dendritic actin protrusive mechanism for advancing on a 2D surface?

Mostov: He sees both in 3D, depending on the cell type and other conditions. He can have amoeboid growth by blocking processes such as MMP production. Depending on how we do our experiments we influence one or the other. We have grown up for 50 years in a 2D world, and we have mostly seen one part of the spectrum of cell motility.

Borisy: You are accepting the conclusion that there is a second mechanism for motility, which is expressed when you challenge cells to move in a third dimension.

Mostov: Sure.

Borisy: Do we have enough evidence for that conclusion? Coming back to Dr Takenawa's results, he has deleted a molecule very important for motility on a 2D surface which also causes a defect in 3D. There is something in common.

Takenawa: We have examined whether WAVE2 regulates collective movement, a type of movement that is observed in epithelial cells. WAVE2 knock down inhibits such movement rather than amoeboid movement.

Firtel: Dictyostelium is a non-mammalian organism, but it has been useful. There are a lot of mutants in *Dictyostelium* that have difficulty moving through 3D and not along 2D surfaces. These are normally mutants that affect cortical tension, such as myosin, Myo2 and gap junctions. As the analysis has got better, things that have been thought to be leading edge functions end up also being associated with the back of the cell and myosin in the sides of the cell. I was wondering whether the effect of WAVE2 on 3D movement might also be associated with an impact of providing an actin matrix for myosin contractitility or for cortical tension. The impact might not be on a different mechanism of movement but on a different molecular mechanism that is required to strengthen the sides of cells as they need to be able to move through it. In the absence of cortical tension, I envisage that the cell tries to move through cells or a matrix and gets squashed. I was wondering whether the WAVE2 phenotype may be involved also in controlling the other aspect of the cell and myosin contractility.

Borisy: Mike, what is your view on the existence of a different mechanism for crawling through 3D matrices.

Sheetz: The evidence we have is that myosin IIB (M2B)$^{-/-}$ mutants move fine in 2D but don't move collagen fibres well in 3D or contract collagen gels (Meshel et al 2005). The GFP–M2B is fully functional and will restore function in the deficient cells in terms of collagen gel contraction and also in terms of the movement of M2B into lamellipodia. The M2B goes into lamellipodia moving on 3D collagen fibres but the lamellipodia on 2D collagen don't contain M2B, yet the cells move fine.

Borisy: So your conclusion is that M2B is important for motility through a 3D matrix.

Sheetz: Yes. We don't know about M2A yet.

Takenawa: Many proteins are involved in total cell movement. We must distinguish their roles. For instance, RhoA is very important for cell movement. But this protein rather acts at the tail end of the cells and develops stress fibres.

Yap: Although Olivier Pertz and Klaus Hahn claim that they see activation at the front of the cells, as they reported at the ASCB meeting in December 2003.

Takenawa: Rho may be inhibited at the leading edge.

Alberts: Using fluorescence resonance energy transfer (FRET), we can see RhoA interacting with effectors at the leading edge. The Hahn lab has similar data demonstrating that RhoA is activated at the leading side of cells using different approaches. The prevailing evidence for that is that Burridge's RhoA/Rho kinase story demonstrating that Rho and Rho kinase is important for retraction of the tail of migrating cells. It is clear that RhoA localized to the leading edge, and we see tons of it there, along with Rho C. Data from Alan

Hall's lab years ago showed RhoA in ruffles. We hypothesize that RhoA and RhoC are interacting with formins and driving actin polymerization there. RhoA has multiple effectors in addition to Rho kinase. RhoA signalling is more complex than simply being in charge of retraction at the tail end.

Borisy: We are talking about signalling networks, and the network of signalling molecules is a swamp. We enter it, and it is very hard to emerge unscathed and unmuddied.

References

Abram CL, Seals DF, Pass I et al 2003 The adaptor protein Fish associates with members of the ADAMs family and localizes to podosomes of Src-transformed cells. J Biol Chem 278:16 844–16 851

Meshel AS, Wei Q, Adelstein RS, Sheetz MP 2005 Basic mechanism of three-dimensional collagen fibre transport by fibroblasts. Nat Cell Biol 7:157–164

Wolf K, Mazo I, Leung H et al 2003 Compensation mechanism in tumor cell migration: mesenchymal-amoeboid transition after blocking of pericellular proteolysis. J Cell Biol 160:267–277

Cytoskeletal networks and pathways involved in endocytosis

David G. Drubin, Marko Kaksonen, Christopher Toret and Yidi Sun

Department of Molecular and Cell Biology, 16 Barker Hall, University of California, Berkeley, CA 94720-3202, USA

Abstract. Until recently, the actin cytoskeleton and the endocytic machinery were thought to operate independently. However, the actin cytoskeleton is an integral part of the cell cortex and there is growing evidence in diverse eukaryotes that F-actin plays a direct role during endocytic internalization. Genetic studies in *Saccharomyces cerevisiae* have demonstrated that Arp2/3-mediated F-actin assembly is required specifically for the internalization step of endocytosis. Using real-time image analysis, we recently defined a pathway for receptor-mediated endocytosis in budding yeast. Many features of this pathway appear to be conserved widely, indicating that principles derived from our studies in yeast will be directly applicable in more complex eukaryotes. We are pursuing our yeast studies using a combined approach involving image analysis, functional genomics, proteomics and biochemistry. These ongoing studies are providing a broader and deeper understanding of the molecular events of endocytosis, of how forces for actin polymerization are harnessed, and of how steps in the pathway are regulated. Our studies in mammalian cells provide evidence that this pathway is conserved in more complex organisms for endocytic and Golgi trafficking events.

2005 Signalling networks in cell shape and motility. Wiley, Chichester (Novartis Foundation Symposium 269) p 35–46

Endocytosis plays a crucial role in regulation of plasma membrane composition and in uptake of molecules from a cell's environment. Several endocytic pathways have been discovered. These pathways differ in the type of cargo endocytosed, in the volume endocytosed and in the underlying machinery. Here we focus on receptor-mediated endocytosis (RME). While it has long been appreciated that RME occurs in close proximity to a cell's actin-rich cortex, only recently has it become clear that the actin cytoskeleton is an active participant in RME (Engqvist-Goldstein & Drubin 2003). In yeast cells, mutants of actin cytoskeleton proteins and the actin inhibitor latrunculin A block RME. In vertebrate cells, inhibitor-based evidence for a role for actin in RME has been less definitive (Fujimoto et al 2000). However, recent studies involving real-time analysis of fluorescently tagged proteins have convincingly demonstrated an endocytic pathway at the plasma membrane in which clathrin appears first, then the GTPase dynamin and then, shortly thereafter, actin (Merrifield et al 2002). Furthermore, RNAi of the highly conserved endocytic protein Hip1R, which binds to clathrin, phosphatidylinositol-4,5-bisphosphate (PIP2) and actin, caused actin filaments, often in the form of tails, to associate stably with clathrin structures at the plasma membrane (Engqvist-Goldstein et al 2004). Fluorescence recovery after photobleaching (FRAP) studies demonstrated that the actin in these structures turns over dynamically. A role for actin in RME is therefore highly conserved across eukaryotes. Still lacking is clarity on functional, regulatory and mechanistic aspects of actin's role in RME.

Certain features of the budding yeast *Saccharomyces cerevisiae* make it well suited for studies of RME. First and foremost are its well-recognized genetics and molecular genetics, which make routine the inactivation of any gene in the genome. Second, numerous reagents and datasets from a variety of genomic and proteomic studies are available. For example, a null mutant of every gene has been constructed, every gene has been tagged with green fluorescent protein (GFP), and for most genes a number of physical and genetic interactions have been identified and gene expression profile data are available (Martin & Drubin 2003). As will be described below, these various tools can help guide studies of yeast cell biology. Also, though small in size, budding yeast are in many ways ideally suited for real-time image-based analyses of RME. GFP and its spectral variants can be fused to any yeast protein in any genetic background by a simple *in vivo* integration procedure, such that the resulting fusion protein is expressed from its own promoter and at normal levels, as the only form of that protein in the cell (Longtine et al 1998). The functionality of such fusions can also be evaluated readily. If mutation of the gene results in a detectable phenotype, the fusion protein can be tested for that phenotype. A final advantage of yeast for studies of cortical actin dynamics is that the only type of actin structure detectable on the surface of a yeast cell is the cortical actin patch, which is the actin-based structure that mediates RME (Sekiya-Kawasaki et al 2003). This lack of complexity in actin-based cell surface structures makes it possible to monitor the dynamics of every actin patch in a given focal plane for minutes on end.

The above-mentioned 'cortical actin patches' are one of three types of filamentous actin structure in yeast (the other two being a network of cytoplasmic cables and a contractile ring that functions during cytokinesis). While the observation that mutations in patch proteins often resulted in endocytic defects suggested that actin participates directly in endocytosis, it might have been the case that cortical actin patches play an indirect role in endocytosis. Detailed analysis of two patch properties, their protein composition and their dynamics, were crucial for the development of a model for how actin patches function directly in endocytosis. Thanks to two-hybrid, immunoprecipitation and affinity isolation studies, including small-scale focused studies and large-scale proteomic studies, numerous components of cortical actin patches have been linked together in an interaction network (Schwikowski et al 2000, Drees et al 2001, Ito et al 2001, Washburn et al 2001, Gavin et al 2002). Most proteins in the network are highly conserved, suggesting that patch function is conserved. Sequence analysis revealed that these patch proteins consist of a mixture of actin cytoskeleton proteins (e.g. the Arp2/3 complex, WASP, actin, cofilin) and endocytic proteins (e.g. epsin, Eps15, adaptors) (Engqvist-Goldstein & Drubin 2003). While the vast majority of these proteins show patch-like localization on the yeast cortex, and while all of the patches tend to cluster in regions of the cell cortex known to be active in exocytosis, co-localization experiments for pairs of patch proteins enigmatically have often revealed at best only partial colocalization of patch proteins.

A typical protein interaction network is shown schematically in Fig. 1a (for a diagram with actual proteins and interactions, please see Engqvist-Goldstein & Drubin 2003). With the growing popularity of 'proteomic' studies, this type of diagram describing myriad protein interactions is becoming increasingly common. Protein interaction networks can be very useful for identifying many if not all of the proteins in a biological pathway. Proteins that appear at nodes in the interaction network, and that are brought into the network by multiple independent experimental approaches, are likely to be bona fide components of the network.

Although, as we will discuss below, the 'network' is best viewed as an abstract concept rather than as an actual depiction of protein status in a cellular system, the extensive complexity of such systems and the extensive overlap in protein–protein interactions impart important functional and regulatory properties to the systems. Such systems often have a property referred to by geneticists as 'robustness', in which damage to one protein has minimal impact on the process because other proteins are present with overlapping interactions and functions (Wagner 2000). This principle appears to apply to many actin cortical patch proteins. For example, four Arp2/3 activators exist in patches. Mutations of the Arp2/3 activating domain of yeast WASP or Eps15-like Pan1p cause hardly any phenotype. However, a more pronounced phenotype is seen when two such mutations are combined in a double mutant (Duncan et al 2001). Thus, when making mutants in components of a network, phenotypes are only observed for those protein functions that are not redundant. Complex networks also provide an opportunity for increased functional and regulatory complexity (Thomas 1993). This is due to combinatorial effects that result from changing the activities or levels of subsets of proteins within the network. Each distinct combination of protein activities may have novel functional or regulatory properties. In sum, the high complexity and redundancy of endocytic and cytoskeletal proteins likely contributes to robustness, efficiency and functional/regulatory diversity and adaptability.

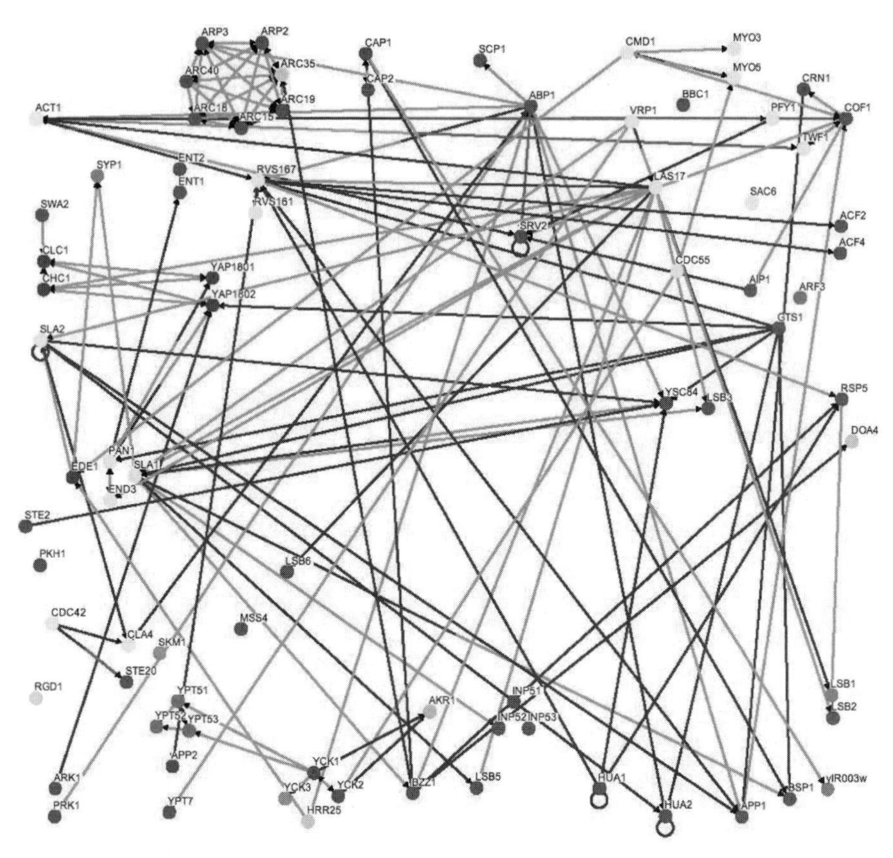

FIG. 1. Endocytic/cortical actin patch protein interaction network. Protein–protein interaction network of 83 genes implicated in endocytosis and actin patch dynamics of *S. cerevisiae*. The network was generated using Osprey 1.2.0 (*http://biodata.mshri.on.ca/osprey/servlet/Index*). The network shows interactions detected by two-hybrid, affinity isolation, purified complex and affinity chromatography approaches as posted in the Yeast GRID database (*http://biodata.mshri.on.ca/yeast_grid/servlet/SearchPage*).

A diagram showing the interaction network of cortical actin patch proteins can provide us with information about the molecular composition of patches, and about connectivity among patch proteins (Fig. 1). But how does an interaction network drawn to summarize protein–protein interactions reflect on the cellular status of proteins in the network? First, some caution in viewing interaction network diagrams is appropriate because proteomic studies vary in quality and the datasets tend to be noisy. Criteria for establishing the validity of interactions depicted in network diagrams varies widely, as do the methods used for finding interactions. A second important point is that a diagrammed interaction network

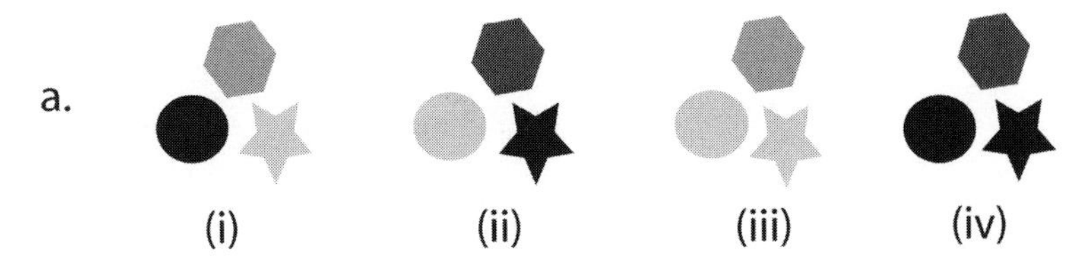

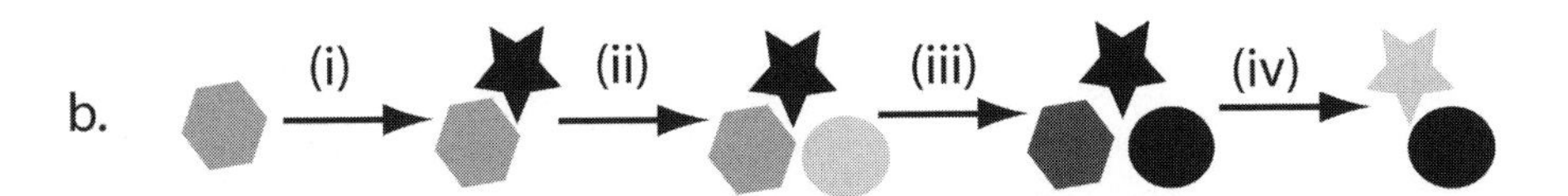

FIG. 2. Protein interaction models. Each shade/shape combination represents a distinct protein. (a) It may be that only certain combinations of protein interactions depicted in diagrams like the one shown in Fig. 1 can occur simultaneously. This may be due to spatial segregation of proteins, mutually exclusive binding interactions, or regulation of the timing of interactions. Therefore, despite the fact that it is possible to depict the proteins in a large interaction network, only certain protein combinations, containing specific subsets of proteins, interact in distinct complexes (i–iv) in cells. These distinct complexes might have distinct functions and/or distinct responses to regulatory signals. (b) In the second model, most interactions depicted in an interaction network diagram occur during a process, but many interactions occur only transiently to impart a certain function upon a process. Here, the interactions occur transiently in a four-step (i–iv) linear pathway like the one we have proposed for RME (Kaksonen et al 2003).

is not something that would likely ever exist in a cell. If all of the interactions depicted in a network diagram were to form at once, a large precipitate would form in the cytoplasm.

If the proteins in the endocytic protein network cannot all interact with each other simultaneously, what then is the interaction network diagram telling us about the interactions and functions of the proteins in the network? One possibility is that not all of the interactions can occur at once because some interactions are regulated, or they are mutually exclusive, or because different proteins are expressed at different times, or at different locations in cells. Thus, complexes containing only a subset of proteins might exist, and different combinations of proteins might combine to create complexes with functional and regulatory differences (Fig. 2a). An alternate possibility is that the majority of the interactions occur every time a process like endocytosis occurs, but the interactions occur sequentially in a highly regulated manner akin to a factory production line (Fig. 2b). For this model, proteins interact sequentially to perform specific functions, and they dissociate once their functions have been provided.

Results and discussion

To determine whether, when and where the different proteins in the cortical actin patch interaction network interact in living cells, we tagged pairs of patch proteins with different coloured fluorescent proteins, and monitored their dynamics at high time resolution (Kaksonen et al 2003). This type of analysis was performed on many pairwise combinations of six patch proteins. In the end, it was determined that these six proteins associate with each other in cells in a manner similar to the production line model (Fig. 2b). When a single patch was observed over time, the composition of that patch underwent a very stereotypical series of changes, with a highly predictable order and timing. Only one type of yeast cortical actin patch exists, but its composition changes in a regular manner with predictable timing from the birth of the patch to its disappearance. There are early patch proteins and late patch proteins in this pathway. The early proteins include Pan1p (an Eps15-like Arp2/3 activator), Las17p (a WASP-like Arp2/3 activator), Sla2p (which binds to actin, clathrin and PIP2) and Sla1p (which can bind to several actin cytoskeleton proteins plus the Ste2p receptor, which is an endocytic cargo protein). Late proteins in the pathway included the Arp2/3 complex, actin and Abp1p (an actin-binding protein that activates the Arp2/3 complex). Importantly, when the late proteins, which include actin and actin assembly regulators, associate with the endocytic complex, the complex starts to move.

There are two forms of patch movement. A slow form of movement occurs first, and is followed by a faster form of movement. We proposed that the slow form of movement, which occurs over distances of 100–200 μm, represents the invagination of the plasma membrane. The faster form of movement, we proposed, represents the movement of the free endocytic vesicle subsequent to the scission event. We further found using the actin inhibitor latrunculin A that both forms of movement depend on the actin cytoskeleton, and we proposed that actin assembly might drive plasma membrane invagination and vesicle scission.

In this pathway, proteins appear at the plasma membrane in a predicable order, with very regular timing. Motility events also occur in a predictable manner, subsequent to the appearance of actin at the patch. Slow motility occurs as actin intensity reaches a maximum, and fast motility occurs at a time when actin intensity is declining. In mutants lacking Sla2p, actin 'tails' become stably associated with endocytic patches, which do not move off the plasma membrane. FRAP studies demonstrated that new actin assembly occurs at the endocytic sites. This result and earlier studies showing that mutants in actin and cofilin that reduce actin dynamics also block endocytosis (Lappalainen et al 1997, Belmont & Drubin 1998), indicate that actin assembly and filament turnover are intimately coupled to endocytic internalization. Each of four Arp2/3 complex activators had a distinct dynamic behaviour, suggesting that each might be responsible for harnessing actin dynamics for a different type of event (e.g. membrane invagination, vesicle scission, vesicle propulsion) during endocytic internalization (Kaksonen et al 2003).

To flesh out the pathway, one can look to the proteomic interaction network to find candidates for additional proteins (Fig. 1). Our analysis of proteins in the network, and of other proteins implicated in endocytosis by phenotypes or sequence homology, identified about 57 candidates in total. These proteins are being tagged so their time of appearance at the plasma membrane can be determined. They are also being mutated to test their functional importance in the pathway. Not surprisingly, given the highly intricate network of interactions (Fig. 1) and the above-mentioned redundancy within the network, we found that most mutants had no detectable phenotype. However, mutants of 13 of the 57 proteins did have detectable phenotypes, and our continued analysis of these 13 mutants promises to reveal the distinct roles of each protein, and to identify distinct stages at which the endocytic internalization process can be arrested.

Conclusions and future directions

In defining a pathway for endocytic internalization, we created an opportunity to better frame the key questions that need to be answered to increase our understanding of the pathway. One important goal is to define the earliest steps in the endocytic pathway, and to identify the initial events that trigger assembly of an endocytic site. Furthermore, we presently know little about how the order and timing of events are determined. Protein kinases, GTPases and lipid-modifying enzymes all are likely regulators. While some candidate regulators have been identified, how their activities are controlled is not known and the roles of these putative regulators are largely untested. We also need to determine when cargo molecules associate with the endocytic machinery. In principle, the endocytic machinery could assemble around cargo molecules once they are targeted for endocytosis. Alternatively, the machinery could assemble first and cargo could then associate with the machinery. Another crucial question is how actin assembly forces are harnessed to invaginate the plasma membrane. Particularly urgent for answering this question is development of an ultrastructural understanding of the organization of patch proteins. Four Arp2/3 activators are patch proteins and the specific contribution of each of these proteins to endocytic internalization needs to be determined. Even more intriguing is the observation that one of the Arp2/3 activators is a class I myosin. In budding yeast, the two class I myosins contain in addition to motor domains, an Arp2/3 activating domain. The roles of the motor and Arp2/3 activating domains need to be determined. Other important questions concern how the endocytic vesicles become uncoated. It appears that the Prk1p and Ark1p kinases are involved, but that begs the question of how these kinases are regulated so uncoating will occur at the appropriate time. Another issue awaiting an answer is how endocytic vesicles are delivered to endosomes, an event that must be coordinated with uncoating. Finally, now that it is becoming increasingly clear that the actin cytoskeleton participates in endocytosis in yeast and mammals, there is a need to test the extent to which what is being learned in yeast will translate to more complex eukaryotes.

References

Belmont LD, Drubin DG 1998 The yeast V159N actin mutant reveals roles for actin dynamics in vivo. J Cell Biol 142:1289–1299

Drees BL, Sundin B, Brazeau E et al 2001 A protein interaction map for cell polarity development. J Cell Biol 154:549–571

Duncan MC, Cope MJ, Goode BL, Wendland B, Drubin DG 2001 Yeast Eps15-like endocytic protein, Pan1p, activates the Arp2/3 complex. Nat Cell Biol 3:687–690

Engqvist-Goldstein AE, Drubin DG 2003 Actin assembly and endocytosis: from yeast to mammals. Annu Rev Cell Dev Biol 19:287–332

Engqvist-Goldstein AE, Zhang CX, Carreno S et al 2004 RNAi-mediated Hip1R silencing results in stable association between the endocytic machinery and the actin assembly machinery. Mol Biol Cell 15:1666–1679

Fujimoto LM, Roth R, Heuser JE, Schmid SL 2000 Actin assembly plays a variable, but not obligatory role in receptor-mediated endocytosis in mammalian cells. Traffic 1:161–171

Gavin AC, Bosche M, Krause R et al 2002 Functional organization of the yeast proteome by systematic analysis of protein complexes. Nature 415:141–147

Ito T, Chiba T, Ozawa R et al 2001 A comprehensive two-hybrid analysis to explore the yeast protein interactome. Proc Natl Acad Sci USA 98:4569–4574

Kaksonen M, Sun Y, Drubin DG 2003 A pathway for association of receptors, adaptors, and actin during endocytic internalization. Cell 115:475–487

Lappalainen P, Fedorov EV, Fedorov AA, Almo SC 1997 Essential functions and actin-binding surfaces of yeast cofilin revealed by systematic mutagenesis. EMBO J 16:5520–5530

Longtine MS, Fares H, Pringle JR 1998 Role of the yeast Gin4p protein kinase in septin assembly and the relationship between septin assembly and septin function. J Cell Biol 143:719–736

Martin AC, Drubin DG 2003 Impact of genome-wide functional analyses on cell biology research. Curr Opin Cell Biol 15:6–13

Merrifield CJ, Feldman ME, Wan L, Almers W 2002 Imaging actin and dynamin recruitment during invagination of single clathrin-coated pits. Nat Cell Biol 4:691–698

Schwikowski B, Uetz P, Fields S 2000 A network of protein-protein interactions in yeast. Nat Biotechnol 18:1257–1261

Sekiya-Kawasaki M, Groen AC, Cope MJTV et al 2003 Dynamic phosphoregulation of the cortical actin cytoskeleton and endocytic machinery revealed by real-time chemical genetic analysis. J Cell Biol 162:765–172

Thomas JH 1993 Thinking about genetic redundancy. Trends Genet 9:395–359

Wagner A 2000 Robustness against mutations in genetic networks of yeast. Nat Genet 24:355–361

Washburn MP, Wolters D, Yates JR, 3rd 2001 Large-scale analysis of the yeast proteome by multidimensional protein identification technology. Nat Biotechnol 19:242–247

DISCUSSION

Vallee: The actin tails are thought to be involved in the budding of bacteria from cells. You have tails responsible for invagination. How do you see that functioning, since the topology doesn't work so well?

Drubin: There are a couple of possibilities. One is that the geometry in the Sla2 mutant gets turned around such that regulation is lost. Normally, the endocytic site has to achieve a certain geometry before there is assembly. But if you have assembly at the plasma membrane and then anchor the endocytic machinery to the actin cytoskeleton, and it continues to assemble at the membrane, you could invoke a model in which this structure pushes off like a rocket blasting off from the surface of the cell. This touches on the most embarrassing void in our understanding: the ultrastructure of these endocytic structures. We have done a detailed study on this by deep etch electron microscopy (EM), unroofing the cells. We can see pyramid structures. Beyond this, we haven't quite put the pieces together. We have decided to invest in trying to reconstitute this system *in vitro* because it has been hard for us to get much structural information by EM, beyond the fact that we see actin coating these cones. It seems that the Arp2/3 complex is at the apex of the cones. Perhaps this model works where it is adhering to membrane and sucking it out.

Borisy: You have the Arp activator at the membrane.

Drubin: Yes. We actually have four Arp activators.

Borisy: Let's be explicit about this, because it is a very interesting question. If you have an Arp activator at the membrane, and you have polymerization starting at the membrane, then presumably the barbed end of the actin is at the membrane. You don't know the polarity yet, but this is what you would be assuming. Is that correct? As the vesicle invaginates, it could be bound to the actin tail, and then polymerization of the membrane could drive the tail backwards, along with whatever is bound to it.

Drubin: Assembly is occurring at the membrane, and the structure is anchored to an actin network. In EM we have seen that the Arp2/3 complex is at the apex of the invagination, but this doesn't completely mesh with the light microscopy. We have seen it with three different antibodies by immuno-EM. We have evidence that this assembly can be perpetuated. The Arp complex is needed to initiate the assembly. If we make mutants in the Arp complex we slow down the process and stretch out the period during which actin is assembling. Once the actin assembles, this treadmilling recurs at almost the same rate as without functional Arp.

Borisy: Is the polymerization at the membrane therefore Arp independent?

Drubin: That's a little bit too strong a statement. Once you generate the ends of filaments, it could all be growth from these barbed filament ends. If we make Arp mutants and look at treadmilling, we see that assembly is occurring at the normal rate, but the network starts to break up. This phenotype correlates with the block of endocytosis when we look at different Arp mutants.

Borisy: After the invagination becomes a vesicle and the vesicle moves away quickly, you are saying that this is an actin-driven process. What is the polarity at that point?

Drubin: This was inspired by work by Fred Chang in *Schizosaccharomyces pombe* and Lisa Pon has a similar story that Patch is moving on cables. She reached the same conclusion as us but has better evidence. In this actin cable system the cables are assembling rapidly (extending at 300 nm/s). Mutants that block cable assembly block that motility. Lisa has some nice evidence that the vesicles and cable move together. It seems that the vesicles are being loaded on to these cables that are being assembled by formins and are moving away from the tip.

Firtel: What is myosin I doing?

Drubin: That is a great question. It appears late in the pathway. It is essentially there when the fission event occurs.

Firtel: I thought myosin I was supposed to be involved in moving WASP on the actin cables, keeping it in the front.

Drubin: Perhaps that is what it is doing. These guys are intimately connected through the WIP protein. WASP and myosin I do not move off the membrane, whereas this other activator, Pan1, moves off the membrane.

Gundersen: I don't understand about comet tails. So comet tails are not involved in your thinking?

Drubin: Yes, they are. This is a transient event and once things start moving there are about 3–5 seconds, and then actin disappears. But in these cases we are seeing long tails that are long lived. One possibility is that the invagination step has been uncoupled in the formation of this actin structure. It could be that the demise of this whole structure is initiated by the invagination of the membrane. There are proteins that get recruited to the structure specifically when the membrane curves. These could have a scaffolding function to bring in negative regulators. Perhaps if you don't get curvature then you have uncoupled the assembly from the invagination of the membrane and this structure just grows out.

Chang: In the Sla2 mutants, Arp2/3 appears to be well away from cell surface, deep inside the cell. In this case, do you think that the Arp2/3 is dissociated from the invagination, or is the invagination itself very deep?

Drubin: I don't know the answer to that. We haven't looked by light or electron microscopy for Arp in the Sla2 mutant cells.

Sheetz: Are actin filaments only associated with this complex in this pathway, and in no other pathway?

Drubin: There are three filamentous actin systems in yeast: the contractile ring, the cable network (nucleated by formins) and patches. We think that there is just one kind of patch. We have tagged 27 proteins in these patches, and every single one falls in a single linear pathway.

Cai: You suggest that these patches are present for endocytosis. But the patches are very well correlated with cell growth and secretion. If you start with an unbudded cell, where it polarizes and where the bud grows is correlated with the patches.

Drubin: This is a common theme in a variety of cells: the endocytic machinery seems to be coupled with the exocytic machinery. For example, in the synapse there are active zones of exocytosis and around that there are active zones of endocytosis to recycle the membranes. We used to think it was a simple cyclical process. Now there is another dimension to this: as a cell becomes polarized there is a polarity structure that assembles on the surface of the cell. Cables are recruited to this. It is a self-perpetuating system where once the cables are associated with this signalling patch, other things are recruited to this region of the surface, which then starts to grow. It makes sense if you are secreting here that you also have active endocytosis here. Also, we now know that endocytosis is largely mediated by these cables. The patches are riding on cables and the endosomes are riding to the pathway on a cable. It seems logical that this system is designed to be efficient by coupling these patches with the cables.

Cai: Are you saying that endocytosis is required for bud growth?

Drubin: No.

Cai: How do the two couple?

Drubin: We don't know the mechanism yet. One of the important questions is how the endocytic machinery is coupled to the formins. There must be a coupling.

Sheetz: There is a physical constraint here: if you don't have exocytosis you can't have endocytosis, because the cell has a finite volume. In plant cells it is quite clear that if you put them in a hypotonic medium they don't endocytose until their membrane tension drops to a certain point, i.e. until they exocytose enough membrane.

Borisy: In a growing system, volume need not be conserved.

Drubin: It is a homeostatic mechanism. Without endocytosis to recover membrane, the surface would expand about five times faster than it does. The rate of growth relies on a balance between endo- and exocytosis.

Alberts: It appears that actin contributes to endocytosis at different steps. If so, are these incremental? Is it possible to use drugs to ask at what point actin polymerization is critical? For example, could you add latrunculin to yeast and determine what the contribution of polymerized actin is at a given point? You could compare this with cytochalasin D. When we have done these types of experiments with epidermal growth factor (EGF) we see very different effects of cytochalasin D and latrunculin when monitoring cells with fluorescent cargos.

Drubin: Sadly, cytochalasin D does nothing in yeast cells, because it gets pumped out. We searched for latrunculin in the first place because we wanted a chemical tool. This was a great advance in itself. It is the only tool we have in yeast for depolymerizing actin.

Peter: Is Cdc42 involved in these patches?

Drubin: To me that is a mysterious connection. You can connect it up through this bifurcating pathway, but it isn't clear what this means mechanistically in terms of physical interactions. Cdc42 is organized differently in the patches. Somehow there is communication among Cdc42, the Rho proteins and formins. There are many unanswered questions.

Control of cell polarity in response to intra- and extracellular signals in budding yeast

Michele Knaus, Philippe Wiget, Yukiko Shimada and Matthias Peter

Institute of Biochemistry, ETH Hönggerberg, 8093 Zürich, Switzerland

Abstract. Budding yeast serves as a powerful genetic model organism for studying the molecular mechanisms of cell polarity in single cells. Like other polarized eukaryotic cells, yeast cells possess polarity programs that regulate where they grow and divide. Establishment of a site of cell polarity may be conceptualized in several stages. First, cells mark a specific location at the cell surface for polarized cell growth and cell division. To define these sites, cells use intrinsic cues present in the cell or landmarks determined by extracellular signals such as morphogens. Second, these landmark proteins then recruit or activate polarity establishment proteins including small GTPases and their regulators. Positive and negative feedback mechanisms are required to transform these site-selection processes into a stable axis of polarity. Finally, these locally activated GTPase modules recruit and activate proteins that organize the actin cytoskeleton and cell growth. In this short review, we describe molecular pathways required to establish oriented cell polarity, and emphasize recent advances in defining positive and negative feedback mechanisms that together may translate an initially weak symmetry-breaking signal into a robust axis of polarity.

2005 Signalling networks in cell shape and motility. Wiley, Chichester (Novartis Foundation Symposium 269) p 47–58

Site-specific activation of Cdc42p is critical for the establishment of cell polarity

The yeast *Saccharomyces cerevisiae* has proven to be an excellent and genetically tractable model organism to study the biochemical pathways leading to the establishment of cell polarity (Fig. 1A, Pruyne & Bretscher 2000). During vegetative growth, polarization of the actin cytoskeleton towards a single position on the cell cortex (incipient bud site) leads to bud formation, while during mating, a pointed projection (shmoo) develops to allow contact and fusion with a cell of the opposite mating type. Central to the initiation of actin

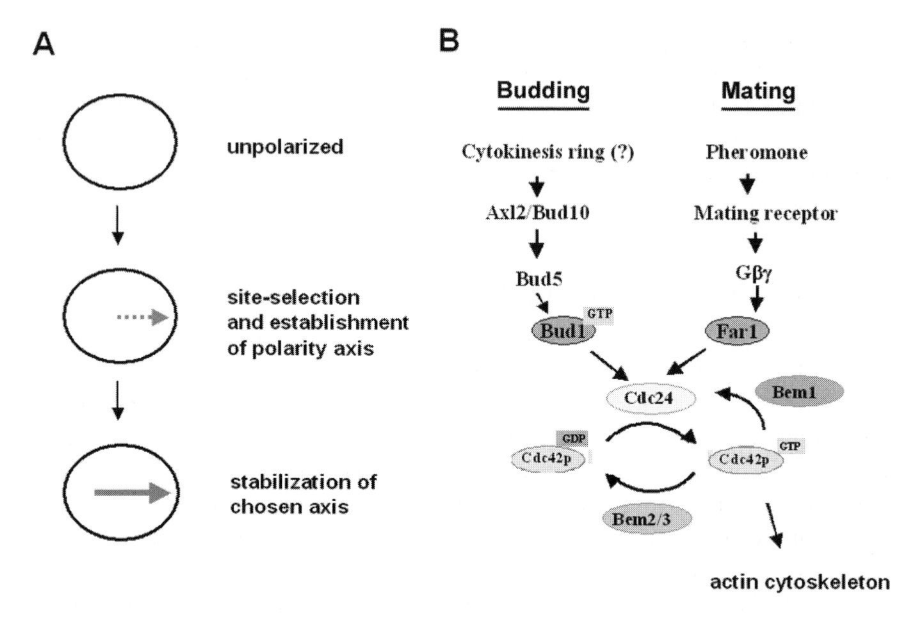

FIG. 1. Pathways required to establish cell polarity during budding and mating. (A) Schematic drawing depicting the different stages involved in cellular polarization. A site of polarization is determined by internal cues or external signals. This initial weak polarity axis then needs to be stabilized by positive and negative feedback mechanisms. (B) Molecular pathways involved in polarity establishment in budding yeast during budding (internal cue) and mating (pheromone gradient). The cortical landmark proteins Axl2/Bud10p or the activated mating receptors recruit the Cdc42p GEF Cdc24p via the small GTPase Rsr1p Bud1p or the scaffold protein Far1p. The adaptor Bem1p is part of a positive feedback loop to amplify Cdc24p activity at the site of polarization. The GAPs Bem2p and Bem3p may spatially restrict Cdc42p activity. Local activation of Cdc42p leads to the polarized assembly of the actin cytoskeleton.

polarization during mating and budding is the local activation of the GTPase Cdc42p (Fig. 1, Johnson 1999). Loss of Cdc42p-function prevents cell polarization, while cells expressing activated Cdc42p self-organize and develop an axis of polarity. Moreover, cells expressing a hyperactive *cdc42* allele establish several axis of polarity, resulting in the formation of multibudded cells (Caviston et al 2002). Together, these results suggest that site-specific activation of Cdc42p is necessary and sufficient to establish an axis of polarity and polarize the actin cytoskeleton. The regulation of Cdc42p involves its sole guanine nucleotide exchange factor (GEF) Cdc24p and several putative GTPase-activating proteins (GAPs) (Bem2p, Bem3p, Rga1p and Rga2p) (Johnson 1999). In addition, Cdc24p is positively regulated by binding to the adaptor protein Bem1p, which is required to maintain Cdc24p at sites of polarized growth (Butty et al 2002).

Finally, Cdc24p is sequestered in the nucleus during the G1-phase of the haploid cell cycle by binding to the adapter Far1p (Toenjes et al 1999, Nern & Arkowitz 2000, Shimada et al 2000). Importantly, Cdc24p localizes to the chosen site of polarization prior to Cdc42p activation, implying that unravelling the mechanisms of site-specific recruitment and subsequent activation of Cdc24p are key to understand the mechanisms of polarity establishment.

Site-specific recruitment of Cdc24p during mating and budding

During mating, cell polarity is determined by a gradient of mating pheromones secreted by the mating partner (Arkowitz 1999, Gulli & Peter 2001). The cells interpret this pheromone gradient and polarize their cytoskeleton in the direction of the highest pheromone concentration (Segall 1993), thereby ensuring efficient fusion of the two mating partners. Pheromones bind to cell-type specific seven-transmembrane receptors (Ste2p or Ste3p), which in turn activate the associated heterotrimeric G protein (α, β, γ encoded by *GPA1*, *STE4* and *STE18*, respectively), by inducing its dissociation into Gα-GTP and G$\beta\gamma$ subunits (Sprague & Thorner 1992, Leberer et al 1997). Activated G$\beta\gamma$ interacts with the ring-finger proteins Ste5p (Elion 2001) and Far1p (Arkowitz 1999, Gulli & Peter 2001). Ste5p functions as a scaffold protein, which together with the PAK-like kinase Ste20p activates the MAP kinase module composed of the MEKK Ste11p, the MEK Ste7p and the MAP kinase Fus3p. In contrast, Far1p interacts with Cdc24p and G$\beta\gamma$ and is required to interpret the pheromone gradient by recruiting Cdc24p to the site of receptor activation. Consistent with this model, mutations that interfere with the binding of Cdc24p and Far1p prevent directed polarization during mating. Conversely, uniform expression of Far1p at the plasma membrane is sufficient to ectopically recruit Cdc24p (Wiget et al 2004). These results suggest that Far1p is required to orient cell growth in a pheromone gradient by recruiting Cdc24p to the site of receptor activation marked by G$\beta\gamma$.

During vegetative growth, the position on the cell cortex where Cdc24p is activated is not random, but is regulated by several gene products in a cell-type dependent manner (Chant 1999). The transmembrane protein Bud10p/Axl2p is required to guide the budding machinery to the correct position in haploid cells, while Bud8p and Bud9p mark the site of polarization in diploid cells. The GTPase Rsr1p/Bud1p is regulated by its GEF Bud5p and its GAP Bud2p, which both localize to the incipient bud site (Park et al 1999). Bud5p co-localizes with the transmembrane protein Axl2p/Bud10p, a component of the budding landmark, and directly interacts with its cytoplasmic tail (Kang et al 2001), suggesting that Bud5p may be recruited to the incipient bud site by binding to Axl2p/Bud10p. Several lines of evidence suggest a crucial role for the small GTPase Rsr1p/Bud1p in the local activation of Cdc24p at bud emergence. Rsr1p/Bud1p was originally isolated as a multicopy suppressor of *cdc24-4* mutant cells (Bender & Pringle 1989), and subsequently was shown to bind directly to Cdc24p in a GTP-dependent manner (Park et al 1997). Moreover, the recruitment of Cdc24p to the proper bud site depends on Rsr1p/Bud1p (Park et al 2002). Finally, cells expressing Cdc24p mutants defective for the interaction with Rsr1p/Bud1p exhibit a random budding pattern (Shimada et al 2004). Together, these results suggest that Rsr1p/Bud1p is an important regulator of Cdc24p, which recruits Cdc24p to the incipient bud site.

Site-specific activation and amplification of Cdc24p during polarization

Available results suggest that Far1p and Rsr1p/Bud1p may play a dual role in the site-specific activation of Cdc24p. First, Far1p and activated Rsr1p/Bud1p recruit Cdc24p to the mating and incipient bud site, respectively (Fig. 1B, Chang & Peter 2003). Second, binding of Far1p and Rsr1p/Bud1p to Cdc24p at the cell cortex may be involved in the initial activation of Cdc24p. Cdc24p is regulated *in vivo* by auto-inhibition by its C-terminal PB1^{Cdc24} domain (Shimada et al 2004), and it is possible that binding of the site-specific activators Rsr1p/Bud1p and Far1p may trigger this process. Interestingly, the mammalian exchange factors Vav and Dbl have previously been shown to be regulated by intramolecular events (Bustelo 2002), suggesting that this mechanism may be generally important to control the activity of GEFs.

After the recruitment and initial activation of Cdc24p, positive feedback loops are needed to enhance the local Cdc24p activity and establish a stable polarity axis (Fig. 1). Interestingly, recent results demonstrate that the adaptor protein Bem1p is involved in this process. Indeed, *bem1Δ* cells fail to efficiently polarize the actin cytoskeleton (Chenevert et al 1994), although they are able to initially recruit Cdc24p to the correct site (Gulli et al 2000). Bem1p interacts with the PB1 auto-inhibitory domain of Cdc24p (Ito et al 2001, Butty et al 2002), and is thought to stabilize the active conformation of Cdc24p (Shimada et al 2004). Moreover, multimerization of Bem1p clusters many active Cdc24p molecules at the polarity site (Irazoqui et al 2003). In summary, the available results suggest the following pathway to ensure specific activation of Cdc24p at sites of polarization (Wiget et al 2004, Shimada et al 2004): Cdc24p is kept inactive by an auto-inhibition mechanism. Activated Rsr1p/Bud1p or Far1p recruit Cdc24p to sites of polarized growth, and this interaction triggers a conformational change leading to initial activation of Cdc24p. In turn, locally produced Cdc42p-GTP recruits Bem1p into the complex (Butty et al 2002, Irazoqui et al 2003), which stabilizes active Cdc24p by preventing its inhibitory domain from looping back and inhibit its GEF activity. Because Bem1p has the ability to form multimers and also contains a PX-domain that is likely to interact with membranes (Itoh & Takenawa 2002), many active Cdc24p-Bem1p complexes may be stably anchored at sites of polarized growth and produce a strong local Cdc42p-GTP signal, which in turn organizes the actin cytoskeleton. Such a two-step mode of activation ensures that Cdc24p is activated predominantly at incipient sites of polarization, and may explain how active Cdc24p is maintained during polarized growth.

Bem2p and Bem3p: Cdc42p-GAPs important to restrict Cdc42p activity to sites of polarized growth?

Most models predict that the establishment of a stable axis of polarity not only requires site-specific activation and amplification of Cdc42p, but also global inhibition of Cdc42p activity at the cortex (Sohrmann & Peter 2003). However, the molecular nature of such global inhibitors in yeast is sill unclear.

Cells lacking such global inhibitory pathways are predicted to polarize simultaneously towards multiple sites. Interestingly, cells lacking Bem2p exhibit a multibudded phenotype and accumulate Cdc24p at multiple sites around the cortex (M. Knaus and M. Peter, unpublished results). Bem2p encodes a GAP, which has been shown *in vitro* to hydrolyse GTP to GDP on Cdc42p and Rho1p. In addition to Bem2p, Bem3p also possesses GAP-activity towards Cdc42p and Rho1p *in vitro* (Zheng et al 1993), and it has been shown that *bem2Δbem3Δ* cells are unviable (Bender & Pringle 1991). Taken together, these data suggest that Bem2p and Bem3p may together restrict Cdc42p activation (Fig. 2). Supporting this model, wild-type cells overexpressing Bem3p are unviable, while inactivation of Bem2p and Bem3p in G1-unpolarized cells results in premature polarization of the actin cytoskeleton (M. Knaus and M. Peter, unpublished results). One possible interpretation of these results is that Bem2p and Bem3p act as global inhibitors to prevent Cdc42p activation during the G1 phase of the cell cycle. How then is this cortical inhibition overcome by polarizing signals? The answer to this question is not clear, but intriguingly Bem2p and Bem3p are rapidly phosphorylated just prior to polarization (M. Knaus and M. Peter, unpublished results), raising the possibility that phosphorylation of the GAPs may inhibit their activity thereby contributing to the activation of Cdc42p. It will be interesting to determine whether phosphorylation of Bem2p and Bem3p may be spatially restricted. For example, during mating, Ste5p locally activates the MAP-kinase Fus3p, which in turn may phosphorylate and thereby inactivate Bem2p/Bem3p specifically at sites of receptor activation (Fig. 2). In this speculative scenario, local inactivation of Bem2p and Bem3p would synergize with site-specific Cdc24p activation and contribute to the asymmetric activation of Cdc42p. Future experiments are aimed at testing this intriguing model.

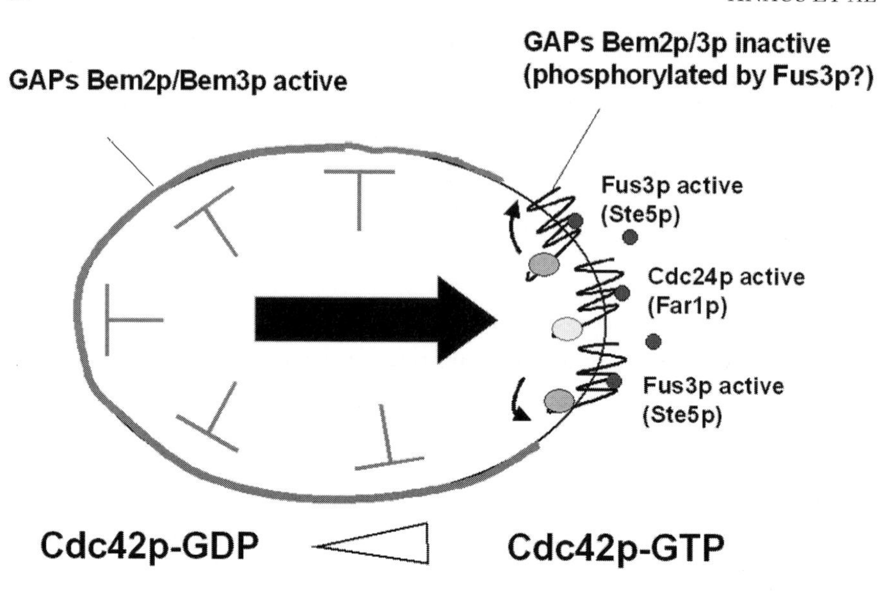

FIG. 2. Positive and negative mechanisms to stabilize the polarity axis during mating. Asymmetric activation of mating receptors in a pheromone gradient (dark grey circles) leads to the dissociation of the heterotrimeric G protein and local accumulation of $G\beta\gamma$, which in turn binds to the scaffold proteins Far1p and Ste5p. Far1p recruits and activates the GEF Cdc24p, which is then stabilized by binding to the adaptor Bem1p. Together this leads to the activation of Cdc42p predominantly at sites of receptor activation. Activated Ste5p assembles the MAP kinase cascade, leading to local activation of the MAP kinase Fus3p. We speculate that Fus3p may locally phosphorylate and thereby inactivate the Cdc42p GAPs Bem2p and Bem3p, thus further enhancing Cdc42p-GTP accumulation at the site of polarized growth.

Additional mechanisms contributing to the establishment of cell polarity

In addition to the mechanisms discussed above, several other positive feedback loops have been described, but their importance for the establishment of cell polarity remains to be determined. In particular, the actin cytoskeleton has been proposed to stabilize cell polarization in many systems (Sohrmann & Peter 2003). Indeed, once the actin cytoskeleton is organized in a polarized manner such that the unbranched actin filaments point towards the site of polarized growth, Myo2p-dependent transport mechanisms target secretion of new membrane material to that chosen site (Pruyne & Bretscher 2000). Thus, polarized secretion may lead to clustering of molecules at the site of polarization. For example, the a-factor receptor Ste3p is rapidly internalized upon ligand binding and at least a fraction of Ste3p is recycled back to the plasma membrane

(Chen & Davis 2000). As a result, it is possible that recycling of Ste3p enhances polarized receptor activation. Likewise, Cdc42p itself may accumulate at the site of polarized growth by polarized secretion. Interestingly, endocytosis and secretion of Cdc42p are important for the self-organizing properties of unpolarized yeast cells exposed to uniform expression of activated Cdc42p (Wedlich-Soldner et al 2003, Irazoqui et al 2003). Future work is expected to unravel the contribution of these mechanisms for the establishment of cell polarity in physiological contexts.

Acknowledgements

We thank members of the Gotta, Barral and Peter groups for stimulating discussion and critical reading of the manuscript. Work in the M.P. laboratory is supported by the Swiss National Science Foundation and the ETH/Zurich.

References

Arkowitz RA 1999 Responding to attraction: chemotaxis and chemotropism in Dictyostelium and yeast. Trends Cell Biol 9:20–27

Bender A, Pringle JR 1989 Multicopy suppression of the cdc24 budding defect in yeast by CDC42 and three newly identified genes including the ras-related gene RSR1. Proc Natl Acad Sci USA 86:9976–9980

Bender A, Pringle JR 1991 Use of a screen for synthetic lethal and multicopy suppresser mutants to identify two new genes involved in morphogenesis in Saccharomyces cerevisiae. Mol Cell Biol 11:295–305

Bustelo XR 2002 Regulation of Vav proteins by intramolecular events. Front Biosci 7:d24–30

Butty AC, Perrinjaquet N, Petit A et al 2002 A positive feedback loop stabilizes the guanine-nucleotide exchange factor Cdc24 at sites of polarization. EMBO J 21:1565–1576

Caviston JP, Tcheperegine SE, Bi E 2002 Singularity in budding: a role for the evolutionarily conserved small GTPase Cdc42p. Proc Natl Acad Sci USA 99:12 185–12 190

Chang F, Peter M 2003 Yeasts make their mark. Nat Cell Biol 5:294–299

Chant J 1999 Cell polarity in yeast. Annu Rev Cell Dev Biol 15:365–391

Chen L, Davis NG 2000 Recycling of the yeast a-factor receptor. J Cell Biol 151:731–738

Chenevert J, Valtz N, Herskowitz I 1994 Identification of genes required for normal pheromone-induced cell polarization in Saccharomyces cerevisiae. Genetics 136:1287–1296

Elion EA 2001 The Ste5p scaffold. J Cell Sci 114:3967–3978

Gulli MP, Peter M 2001 Temporal and spatial regulation of Rho-type guanine-nucleotide exchange factors: the yeast perspective. Genes Dev 15:365–379

Gulli MP, Jaquenoud M, Shimada Y, Niederhauser G, Wiget P, Peter M 2000 Phosphorylation of the Cdc42 exchange factor Cdc24 by the PAK-like kinase Cla4 may regulate polarized growth in yeast. Mol Cell 6:1155–1167

Ito T, Matsui Y, Ago T, Ota K, Sumimoto H 2001 Novel modular domain PB1 recognizes PC motif to mediate functional protein-protein interactions. EMBO J 20:3938–3946

Itoh T, Takenawa T 2002 Phosphoinositide-binding domains: functional units for temporal and spatial regulation of intracellular signalling. Cell Signal 14:733–743

Irazoqui JE, Gladfelter AS, Lew DJ 2003 Scaffold-mediated symmetry breaking by Cdc42p. Nat Cell Biol 5:1062–1070

Johnson DI 1999 Cdc42: An essential Rho-type GTPase controlling eukaryotic cell polarity. Microbiol Mol Biol Rev 63:54–105

Kang PJ, Sanson A, Lee B, Park HO 2001 A GDP/GTP exchange factor involved in linking a spatial landmark to cell polarity. Science 292:1376–1378
Leberer E, Thomas DY, Whiteway M 1997 Pheromone signalling and polarized morphogenesis in yeast. Curr Opin Genet Dev 7:59–66
Nern A, Arkowitz RA 2000 Nucleocytoplasmic shuttling of the Cdc42p exchange factor Cdc24p. J Cell Biol 148:1115–1122
Park HO, Bi E, Pringle JR, Herskowitz I 1997 Two active states of the Ras-related Bud1/Rsr1 protein bind to different effectors to determine yeast cell polarity. Proc Natl Acad Sci USA 94:4463–4468
Park HO, Sanson A, Herskowitz I 1999 Localization of Bud2p, a GTPase-activating protein necessary for programming cell polarity in yeast to the presumptive bud site. Genes Dev 13:1912–1917
Park HO, Kang PJ, Rachfal AW 2002 Localization of the Rsr1/Bud1 GTPase involved in selection of a proper growth site in yeast. J Biol Chem 277:26 721–26 724
Pruyne D, Bretscher A 2000 Polarization of cell growth in yeast. J Cell Sci 113:571–585
Segall JE 1993 Polarization of yeast cells in spatial gradients of alpha mating factor. Proc Natl Acad Sci USA 90:8332–8336
Shimada Y, Gulli MP, Peter M 2000 Nuclear sequestration of the exchange factor Cdc24 by Far1 regulates cell polarity during yeast mating. Nat Cell Biol 2:117–124
Shimada Y, Wiget P, Gulli M-P, Bi E, Peter M 2004 The nucleotide exchange factor Cdc24p may be regulated by auto-inhibition. EMBO J 23:1051–1062
Sohrmann M, Peter M 2003 Polarizing without a c(l)ue. Trends Cell Biol 13:526–533
Sprague GF, Thorner JW 1992 Pheromone response and signal transduction during the mating process of Saccharomyces cerevisiae. In Jones EW, Pringle JR, Broach JR (eds) The molecular and cellular biology of the yeast Saccharomyces. Cold Spring Harbor Laboratory Press, Cold Spring Harbor, p 657–744
Toenjes KA, Sawyer MM, Johnson DI 1999 The guanine-nucleotide-exchange factor Cdc24p is targeted to the nucleus and polarized growth sites. Curr Biol 9:1183–1186
Wedlich-Soldner R, Altschuler S, Wu L, Li R 2003 Spontaneous cell polarization through actomyosin-based delivery of the Cdc42 GTPase. Science 299:1231–1235
Wiget P, Shimada Y, Butty AC, Bi E, Peter M 2004 Site-specific regulation of the GEF Cdc24p by the scaffold protein Far1p during yeast mating. EMBO J 23:1063–1074
Zheng Y, Hart MJ, Shinjo K, Evans T, Bender A, Cerione RA 1993 Biochemical comparisons of the Saccharomyces cerevisiae Bem2 and Bem3 proteins. Delineation of a limit Cdc42 GTPase-activating protein domain. J Biol Chem 25:24 629–24 634

DISCUSSION

Surana: How do cells know where to initiate polarization? Normally, the bud-site should be assembled close to the previous cytokinesis site. Are there biochemical cues in the membrane at the site where formins eventually localize?

Peter: I don't know how the cytokinesis ring or furrow is positioned. The connection is not clear, to my knowledge.

Borisy: This is key to the whole understanding of polarization. Everything begins in your modelling with the marking of the site. This site then recruits other molecules and then through a network of positive and negative regulation you have your axis, but first you must begin with marking the site. There is the issue of marking the site both for internal and external cues. For internal cues you infer a relationship with cytokinesis.

Peter: There is a question mark over this.

Borisy: I thought you were going to relate polarity to the bud scar. Is this part of the marking process?

Peter: I don't think the bud scar is involved in marking. One could say that the bud scar is excluding marking because the proteins can't assemble there. I don't know how Axl2/Bud10 is localized, and this is an important question.

Borisy: With regard to the mating process you showed when you inactivate with stabilizer, something happens which in a sense marks the site. Do you know the mechanism of this? Why do molecules come together to one site and not multiple sites around the cell?

Peter: We don't understand how this initial shallow gradient is translated into a stable gradient. Positive and negative feedback mechanisms are likely involved to strengthen a chosen site, and perhaps to prevent assembly of additional sites.

Borisy: This is similar to a problem that Rick Firtel is going to discuss later in the meeting.

Peter: Yes. What is different about yeast from *Dictyostelium* is that while in both systems the receptors are initially uniformly distributed over the plasma membrane, in the yeast cell the receptors will later cluster at the site of polarization. The mechanisms of this clustering are not understood.

Firtel: The initial problem is essentially the same. If you have receptors all around the cell and a shallow gradient, how do you initiate and stabilize the activation at the site closest to the highest gradient? Do you have any ideas? The cell is bathed in pheromone, but you activate at one site which is predominantly closest to the highest concentration of pheromone. But there is not much difference between the front and back of the cell.

Peter: It is estimated to be about a 1% difference in gradient, which is nothing.

Firtel: Do you have any sense of what the first response is? How do you identify that site?

Peter: We don't have any clever insights. What is remarkable is that the receptor gets internalized rapidly and then it is unclear whether it is then resecreted through the actin cytoskeleton to the site where the polarization signal comes in. The actin-dependent recycling of the receptor may actually be part of such an amplification loop. We have tried to look at the importance of recycling in this system and have not been able to show a critical active role of this.

Borisy: We need to distinguish between two processes here: one process is the polarization of each of the mating cells, and the other is the orientation of the two mating cells towards each other. You described a mutant that polarized but failed to orient. Is that correct?

Peter: Yes. If you analyse mating by real-time microscopy, the first polarization event that we are able to detect is towards the mating partner. At least, the first time we can see sufficient green fluorescent protein (GFP) signal is when the cells are correctly oriented towards the mating partner. Thus, we don't believe that cells first polarize randomly and then correct. However, the polarity axis in a pheromone gradient is dynamic, and the cells have the possibility to correct.

Borisy: Have you decoupled polarization and orientation?

Peter: If you have a mutant that lacks Far1 and then expose this to pheromone, the cells polarize in the wrong direction. They then follow the cortical cue that they would normally follow during budding. They can't orient with respect to the mating partner, but they polarize in the direction of the cytokinesis signal. There it is uncoupled.

Nelson: Can you say a bit more about the Drts? Are they regulated during the cell cycle and do they affect Cdc42 activation?

Peter: I wish I could tell you more about the Drts. What we know is that they seem to be excluded by a Cdc42-dependent mechanism. If we express an activated Cdc42 allele, the Drts are removed from the cortex. Thus, there is a connection to Cdc42, but we don't know what they are doing molecularly. We are intrigued by the localization and from the genetics we believe they have to be involved.

Firtel: Are there any hints from their domain structure?

Peter: They have a domain that is linked to cAMP. We haven't seen an effect of cAMP, though.

Nelson: If you overexpress Cdc42, do they go away from other sites of the membrane?

Peter: If we overexpress Cdc42-GTP then they get off the membrane. How that happens we don't know.

Drubin: What is the effect of loss-of-function?

Peter: It is a mild phenotype. They are a bit bigger but it is nothing dramatic.

Gundersen: If you overexpress Cdc42 you don't get multiple buds, just a bigger cell. Why don't you get more buds?

Nelson: Several researchers have shown that you get multiple buds (Richman & Johnson 2000, Caviston et al 2001, Wedlich-Soldner et al 2003).

Peter: What Rong Li's laboratory has shown in their synchronized system that cells overexpressing Cdc42-GTP still polarize towards discrete sites: sometimes they are multiple, but they are discrete sites. This self-organization mechanism is dependent on endocytosis and secretion, so there is some trafficking involved.

Nelson: Your colleague at Penn showed that a year earlier than her.

Balasubramanian: You have shown nice data on polarizing the cell in G1 while expressing dominant-negative Bem3. Are these Cdc24-independent events? Would you expect them to be?

Peter: We haven't looked, although I would expect them to be.

Balasubramanian: Are these buds at random sites? Do they ignore the previous cues?

Peter: We don't know. The reason is a technical one, because the cells we do this work with exhibit a random budding pattern.

Chang: It has been suggested that secretion might help to establish this positive feedback loop. In particular, Cdc42 might be delivered on secretory vesicles. How about Cdc24? Do your factors such as Bem2 and 3 affect secretion in that manner?

Peter: We have never seen Cdc24 on vesicles involved in any trafficking events. However, the Bems seem to use very interesting trafficking pathways and are secreted to sites in a polarized way. It is also not clear whether the Clns are uniformly inactivated or only in specialized areas. Fred Cross has detected Cln2 I believe also at sites of polarized growth. All we can say currently is that inactivation of the GAPs seems to be important for bud emergence.

Sheetz: In terms of getting a signal out of a background, there are two ways of doing this. One is by spatial integration; the other is by temporal integration. Have you looked into this issue? For example, have you tried to bring another cell close to the stationary cell multiple times to see if multiple contacts or the time of contact is most important?

Peter: We haven't tried triggering from different directions and moving things around because this is very hard to do in yeast and the response is very slow. There is another mechanism that I should have mentioned. I have argued that Bem1 will bind and stabilize Cdc24 while other people have shown that Bem1 can make multimers and oligimerize. You could start with one of these molecules and amplify by scaffolding this together. This would make a highly concentrated Cdc24 by bringing things together.

Kaibuchi: Where is Bem3 phosphorylated by Cdk?

Peter: That is a good question. At the moment we don't know, but Fred Cross has seen polarized localization of Cln2. It may thus be that this inactivation only occurs in a specialized area of the cortex, and thus contributes to the specification of the incipient bud site. The GAPs are active everywhere prior to bud emergence because Cln1 and 2 are inactive. After CDK activation, the GAPs may be completely inactivated or shut down only in a special region.

Kaibuchi: If you make some mutations in Bem3 to make it constitutively active, what happens to the cell?

Peter: Non-phosphorylatable mutants are essentially constitutively active. Then the cells cannot polarize.

Cai: At some point in the cell cycle the polarity has to be destroyed. How?

Peter: That is a good question. When this Bem1/Cdc24 complex is at the tip of the cell it remains active. We believe that the Pak-like kinase Cla4 somehow dissociates Bem1 and Cdc24. Later in the cell cycle we see Cdc24 coming off the cell membrane. We think the mechanism underlying this is activation of the Pak-like kinase, which is an effector Cdc42. It's a sort of negative feedback.

Cai: Is there experimental evidence for this?

Peter: Yes. The piece of evidence missing, though, is that we haven't shown that non-phosphorylatable mutants of Cdc24 actually result in the formation of elongated buds. Thus, we don't know whether the phosphorylation is direct. But if you delete the Pak-like kinase alone you get long buds.

Cai: Which Pak kinase?

Peter: Cla4. It is clear that Cdc24 comes off the tip. The mechanism has to do with the Pak-like kinase Cla4. But there are some conflicting results here about whether there is direct phosphorylation of Cdc24 or not.

Borisy: We know this beautiful network of downstream regulation, with almost all the steps from the initial one, but we don't know the initial one. This is the elusive mystery.

Peter: We work more on the mating side. There we believe it is the G$\beta\gamma$ with the receptor. But on the budding side, you are right.

Borisy: Even on the mating side it seems like you don't yet have an explanation for how this shallow gradient is amplified, and how this results in specifying an axis in a certain direction. We don't know how the two axes develop so that they become oriented to each other. Each cell must make an axis and orient it.

Peter: That is correct.

References

Caviston JP, Tcheperegine SE, Bi E 2001 Singularity in budding: a role for the evolutionarily conserved small GTPase Cdc42p. Proc Natl Acad Sci USA 99:12 185–12 190

Richman TJ, Johnson DI 2000 Saccharomyces cerevisiae Cdc42p GTPase is involved in preventing the recurrence of bud emergence during the cell cycle. Mol Cell Biol 20:8548–8559

Wedlich-Soldner R, Altsschuler S, Wu L, Li R 2003 Spontaneous cell polarization through actomyosin-based delivery of the Cdc42 GTPase. Science 299:1231–1235

Regulation of actin assembly by microtubules in fission yeast cell polarity

Fred Chang, Becket Feierbach[1] and Sophie Martin

Department of Microbiology, Columbia University College of Physicians and Surgeons, 701 W. 168th Street, New York, NY 10032, USA

Abstract. It has been speculated that microtubule plus ends function to regulate the actin cytoskeleton in processes such as cytokinesis, cell polarization and cell migration. In the fission yeast *Schizosaccharomyces pombe*, interphase microtubules regulate cell polarity through proteins such as tea1p, a kelch repeat protein, and for3p, a formin that nucleates actin cable assembly at cell tips. Here, we review recent progress on understanding tea1p regulation and function. Microtubules may govern the localization of tea1p by transporting it on the plus ends of microtubules and depositing it directly onto the cell tip when the microtubule catastrophes. The interaction of tea1p with the CLIP170 protein tip1p is responsible for its localization at growing microtubule plus ends. Tea1p may regulate cell polarity by associating with large 'polarisome' complexes that include for3p. For3p is present at both cell tips, but is not on the microtubules. Tea1p is needed to localize the formin to establish polarized cell growth at cell tips that have not grown previously. These studies begin to elucidate a molecular pathway for how microtubules contribute to the proper spatial regulation of actin assembly and polarized cell growth.

2005 Signalling networks in cell shape and motility. Wiley, Chichester (Novartis Foundation Symposium 269) p 59–72

It is becoming increasingly apparent that the microtubule (MT) and actin cytoskeletons interact in order to carry out their functions (Goode et al 2000, Small et al 2002, Rodriguez et al 2003). In cell migration, MTs are required for proper motility of some cell types, and in directionality in other cell types (Tanaka & Kirschner 1995, Dent & Gertler 2003). It has been speculated that MTs may regulate the assembly of actin at specified locations. One example is in animal cell cytokinesis, where MT plus ends emanating from the mitotic spindle may induce the formation of a contractile ring at the plasma membrane (Canman et al 2003).

[1]Current address: Princeton University, Molecular and Cellular Biology Department, Princeton, NJ 08544, USA.

The fission yeast *Schizosaccharomyces pombe* serves as a model cell type to study MT–actin interactions that regulate cell polarity (Chang 2001, Chang & Peter 2003). Their simplicity, regular cell size and shape make these genetically tractable cells ideal for studies on cell morphogenesis (Snell & Nurse 1993). They are being increasingly used for cell biological studies, and have proved to be amenable for microscopic image analysis. Not only can these cells be manipulated using powerful genetic tools, but recent progress has also been made in developing tools of physical manipulation, such as laser microsurgery, which promise to make *S. pombe* an even more powerful model system for studying conserved cell biological processes.

S. pombe cells grow as rod shapes by cell tip extension. During a cell cycle (of about 2 hours), they grow from 7 μm to 14 μm. The pattern of growth is regulated: after cell division, cells initially only grow from the 'old' end (the end of the cell in the previous cell cycle) and grow in a unipolar manner. At some point in G2 phase, they initiate growth at the new end (the previous septation site), in a process termed New End Take Off (NETO), so that they grow in a bipolar manner (Mitchison & Nurse 1985). Cell growth largely ceases during mitosis and resumes after cell division.

The actin cytoskeleton regulates cell growth. During interphase, actin is organized into actin cables and actin patches. Actin patches are thought to be involved in endocytosis and may be actin filaments organized around endocytic vesicles (Kaksonen et al 2003). It is unlikely that they actively carry materials to the growing cell tips, as previously thought, as they move away from the sites of growth (Pelham & Chang 2001). Actin cables act as tracks for movement and localization of secretory vesicles (Bretscher 2003). These secretory vesicles may contribute to cell growth by delivering enzymes that remodel the cell wall and add new membrane at cell tips. Although actin cables are essential for polarized cell growth in budding yeast, in fission yeast, mutants lacking actin cables (e.g. for3 mutants) still polarize, albeit not normally, and are viable (Feierbach & Chang 2001, Nakano et al 2002). Thus, additional factors may contribute to polarized growth.

As in nerve cells, MTs are not essential for polarized growth in *S. pombe*, but regulate directionality. Mutants with disrupted MTs commonly exhibit bent or branched cell shapes, suggesting that they cannot position or maintain their 'growth' zones properly with respect to the rest of the cell (Toda et al 1983, Sawin & Nurse 1998). Interphase MTs are organized in anti-parallel bundles that are organized from interphase MT organizing centres (iMTOCs) situated on the medial nucleus. These MTs are highly dynamic, as they continually grow out from the MTOCs at the medial nucleus to the cell tip, contact the cell tip for approximately 100 s, and then catastrophe back to the MTOC before growing back again (Brunner & Nurse 2000, Drummond & Cross 2000, Tran et al 2001).

Thus, the cell tips are constantly 'probed' by MT plus ends. Importantly, MT catastrophe is restricted to the regions at the cell tips; this regulation orients the MT bundles along the long axis of the cell, and as described below, allows for the deposition of factors selectively at the cell tips. This spatial regulation of MT catastrophe may be one of the underlying mechanisms ultimately responsible for cell shape control. Numerous factors may regulate MT catastrophe. *In vitro* experiments with pure MTs suggest that simple force of a MT hitting a barrier can cause catastrophe (Janson et al 2003). In this model, the cell tip may selectively contain 'sticky' zones that anchor the MT plus end so that MTs plus ends can slide along the sides of the cell, but get trapped at the cell tips, and their continued polymerization then produces compression stall forces that cause MT catastrophe. Protein factors, such as motors, may also contribute to MT catastrophe regulation (West et al 2001), but their precise roles are not yet clear.

To understand morphogenesis, several genetic screens for cell shape mutants have been performed (Snell & Nurse 1993, Verde et al 1995, Snaith & Sawin 2003). Most of these involved direct microscopic screening of mutated yeast colonies, as most of these genes are not essential. Three common cell shape phenotypes were noted: rounded cells (orbs), bent (bananas), and branched (T-shapes, or tea). The orbs are thought to encode 'general polarity' factors, and many of these are signal transduction proteins and cell wall regulators (Verde et al 1995). The ban and tea mutants resembled tubulin mutants, and not surprisingly, many of these mutants were in tubulin or MT organizing proteins, and exhibited abnormal or absent interphase MTs (Yaffe et al 1996, Radcliffe et al 1998, Vardy & Toda 2000).

The *tea1* mutants have a more unusual constellation of phenotypes (Snell & Nurse 1994, Mata & Nurse 1997). They have polarity phenotypes similar to MT mutants, but MTs are largely normal (although there is a slight but measurable effect in some cells). Cloning of *tea1+* showed that it was a kelch-repeat protein

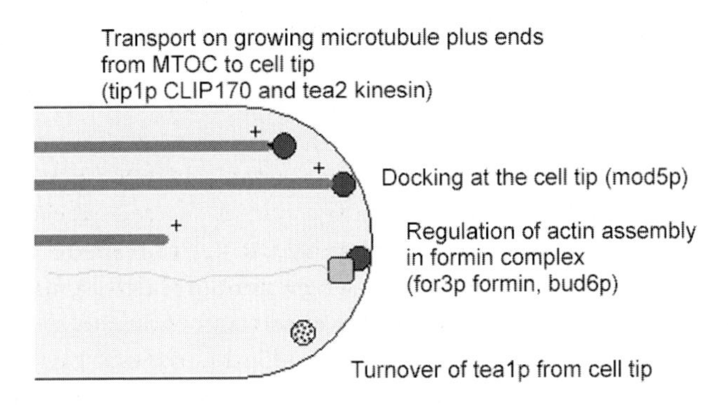

FIG. 1. Regulation of cell polarity at the cell tip in fission yeast: the tea1 cycle.

(Mata & Nurse 1997). Kelch repeats form a β-propeller structure and serve as protein–protein interactions (Adams et al 2000), but this motif was not so informative as to the molecular function of tea1p at the time. Its localization revealed a highly significant pattern: tea1p is located on the MT ends and in dots at both growing and non-growing cell tips and at the septum (Mata & Nurse 1997, Behrens & Nurse 2002). These significant initial findings suggested that tea1p is a potential link between MTs and cell polarity mechanisms, and thus has served as a focus of research to dissect this cellular process. In the paragraphs below, we will review aspects of the 'tea1p cycle' from MT transport to regulation of cell polarity (Fig. 1).

Microtubule transport of tea1p

Time lapse imaging reveals that tea1p localizes in a dot on the growing plus ends of MTs (Behrens & Nurse 2002, Feierbach et al 2004). Tea1p dots are seen travelling from the middle of the cell, just when a new MT is starting to grow, and are transported to the cell tip. Tea1p is generally not on the MT when it shrinks back. Tea1p dots are also seen in the interior of MT bundles; these may represent MT plus ends inside the MT bundle, but also may be tea1p particles moving along the length of the MT. Tea1p depends on other proteins for its localization to MT plus ends. The two best characterized are tip1p, a CLIP170 homologue, and tea2p, a Kip2 class of kinesin, which are also concentrated at MT plus ends (Brunner & Nurse 2000, Browning et al 2000). A number of documented pairwise interactions suggest these proteins form a complex that also includes the EB1 homologue mal3p (Browning et al 2003, Busch & Brunner 2004, Busch et al 2004). For instance, tip1p interacts with tea1p and is required for tea1p accumulation at the cell tip (Feierbach et al 2004). The kinesin tea2p may move both tip1p and tea1p to the MT plus end (Browning et al 2003). In the absence of tip1p or tea2p, tea1 is localized to multiple dots all along the MT. In *tip1* mutants, these tea1p dots on the MTs are motile and exhibit a range of movements in both directions on the MT. Thus, the connections of tea1p to the MT are likely to be complex and may depend on multiple MT motors.

Docking of tea1p at the cell tip

When the MT contacts the cell tip, tea1p dots are transferred from the MT plus end to the cell cortex. These 'docking' events, where a dot of tea1p is left after the MT has shrunk back, have now been directly observed by time lapse microscopy (Feierbach et al 2004). A 'docking' protein mod5p has recently been discovered (Snaith & Sawin 2003). In *mod5* mutants, tea1p does not bind to the cell cortex, but appears (at least in some cases) to stay on the shrinking MT. Mod5p is a prenylated protein that may directly bind to membranes. Interestingly, mod5p distribution is dependent on tea1p, so that their localizations are co-dependent. However, it is not known if mod5p directly binds to tea1p. This docking process presents a potentially interesting mechanism where tea1p has to 'let go' of the MT in response to a higher affinity interaction with the cortex.

Tea1p and regulation of formins

Tea1p regulates cell polarity at the cell tip. Recent studies have identified polarity factors that interact with tea1p presumably at the cell tip. Two of the best understood are bud6p and the formin for3p (Glynn et al 2001, Feierbach & Chang 2001). Formins are a growing family of proteins implicated in cytoskeletal organization and directly nucleate actin filaments through the conserved FH2 domain (Wallar & Alberts 2003). *S. pombe* has three formins, which each organize a distinct actin structure. For3p is required for actin cable formation needed for cell polarity (Feierbach & Chang 2001, Nakano et al 2002). *for3*Δ mutants exhibit altered cell shape and an interesting asymmetrical cell division in which the daughter cells grow in different ways to form cells of different cell shape. For3p is localized to both cell tips and the contractile ring. In contrast to mammalian formins, for3p does not associate with MT nor has any measurable affect on MT dynamics. Bud6p is a formin cofactor that binds to monomeric actin, affects its nucleotide turnover and helps to stimulate formin to assemble actin (Moseley et al 2003). The formin-binding region of bud6p is similar to the region in Rho kinase, although the significance of this region in animal cells is not yet known. The first hint that these proteins had anything to do with tea1p was the phenotype of *bud6* mutants. *Bud6* mutants, like *tea1* mutants, were monopolar, and most strikingly, in a certain background (*cdc11* mutant), formed branches just like a *tea1* mutant (Glynn et al 2001). This phenotype prompted us to investigate the possible interaction of these proteins.

Tea1p interacts with for3p and bud6p in large complexes (45S and 75S) (Feierbach et al 2004). Co-immunoprecipitations, two hybrid assay and altered behaviour of the complexes in mutant strains all showed that these proteins associate together. We speculate that these large complexes represent 'polarisome' complexes that regulate general cell polarity.

Tea1p appears to regulate the localization of for3p and bud6p (Feierbach et al 2004). In *tea1*Δ mutants, for3p and bud6p are only present on one cell tip, and therefore are missing from the second tip, which is always one that has never grown before. Consistent with this effect on formins, actin cables appear to grow from only one cell tip in *tea1*Δ mutants. One simple model is that tea1p may recruit the polarisome to naive cell tips (cell tips that have not grown in a previous cell cycle). However, especially as there are many complex feedback loops, it is not yet clear if tea1p effects are directly on the polarisome, or more indirect.

Tea1p must have functions in addition to regulating formins. Although the single mutants of three genes can still polarize, the triple *for3Δbud6Δtea1Δ* mutant is round and very slow growing (Feierbach et al 2004). These results suggest that these proteins have overlapping functions in cell polarity within a common protein complex. As tea1p has function in the complete absence of actin cables and formin, it may contribute to other aspects of cell polarity such as secretion.

Recently, in order to identify other components of the polarisome, we have identified tea1p-interacting proteins using a biochemical approach (Martin et al 2005). One of these proteins, tea4p, appears to be a functional link between tea1p and for3p. For instance, tea1p and for3p do not appear to bind directly to each other, but tea4p binds directly to both *in vitro*. In particular, tea4p binds to the FH3 domain of for3p, which regulates its localization. The localization and phenotype of tea4p are very similar to those of tea1p. In a *tea4* mutant, tea1p and for3p are disassociated from each other. Overexpression of tea4p stimulates formins to fill up the cell with actin cables. Tea4 is thus a protein at MT plus ends that directly interacts and regulates the formin at the cell cortex.

Conclusion

In summary, the study of cell polarity in *S. pombe* provides an opportunity to probe MT–actin interactions in a simple eukaryotic cell. Our studies are beginning to define an unexpectedly direct molecular pathway for how MT plus ends stimulate formin-dependent actin assembly. As most of these proteins are conserved, one future challenge will be to compare how similar proteins may function together in animal cells in processes such as cell migration and cytokinesis.

References

Adams J, Kelso R, Cooley L 2000 The kelch repeat superfamily of proteins: propellers of cell function. Trends Cell Biol 10:17–24

Behrens R, Nurse P 2002 Roles of fission yeast tea1p in the localization of polarity factors and in organizing the microtubular cytoskeleton. J Cell Biol 157:783–793

Bretscher A 2003 Polarized growth and organelle segregation in yeast: the tracks, motors, and receptors. J Cell Biol 160:811–816

Browning H, Hackney DD, Nurse P 2003 Targeted movement of cell end factors in fission yeast. Nat Cell Biol 5:812–818

Browning H, Hayles J, Mata J et al 2000 Tea2p is a kinesin-like protein required to generate polarized growth in fission yeast. J Cell Biol 151:15–28

Brunner D, Nurse P 2000 CLIP170-like tip1p spatially organizes microtubular dynamics in fission yeast. Cell 102:695–704

Busch KE, Brunner D 2004 The microtubule plus end-tracking proteins mal3p and tip1p cooperate for cell-end targeting of interphase microtubules. Curr Biol 14:548–159

Busch KE, Hayles J, Nurse P, Brunner D 2004 Tea2p kinesin is involved in spatial microtubule organization by transporting tip1p on microtubules. Dev Cell 6:831–843

Canman JC, Cameron LA, Maddox PS et al 2003 Determining the position of the cell division plane. Nature 424:1074–1078

Chang F 2001 Establishment of a cellular axis in fission yeast. Trends Genet 17:273–278

Chang F, Peter M 2003 Yeasts make their mark. Nat Cell Biol 5:294–299

Dent EW, Gertler FB 2003 Cytoskeletal dynamics and transport in growth cone motility and axon guidance. Neuron 40:209–227

Drummond DR, Cross RA 2000 Dynamics of interphase microtubules in *Schizosaccharomyces pombe*. Curr Biol 10:766–775

Feierbach B, Chang F 2001 Roles of the fission yeast formin for3p in cell polarity, actin cable formation and symmetric cell division. Curr Biol 11:1656–1665

Feierbach B, Verde F, Chang F 2004 Regulation of a formin complex by the microtubule plus end protein tea1p. J Cell Biol 165:697–707

Glynn JM, Lustig RJ, Berlin A, Chang F 2001 Role of bud6p and tea1p in the interaction between actin and microtubules for the establishment of cell polarity in fission yeast. Curr Biol 11:836–845

Goode BL, Drubin DG, Barnes G 2000 Functional cooperation between the microtubule and actin cytoskeletons. Curr Opin Cell Biol 12:63–71

Janson ME, de Dood ME, Dogterom M 2003 Dynamic instability of microtubules is regulated by force. J Cell Biol 161:1029–1034

Kaksonen M, Sun Y, Drubin DG 2003 A pathway for association of receptors, adaptors, and actin during endocytic internalization. Cell 115:475–487

Martin SG, McDonald WH, Yates JR, Chang F 2005 Tea4p links microtubule plus ends with the formin for3p in the establishment of cell polarity. Dev Cell 8:479–491

Mata J, Nurse P 1997 tea1 and the microtubular cytoskeleton are important for generating global spatial order within the fission yeast cell. Cell 89:939–949

Mitchison JM, Nurse P 1985 Growth in cell length in the fission yeast *Schizosaccharomyces pombe*. J Cell Sci 75:357–376

Moseley JB, Sagot I, Manning AL et al 2003 A conserved mechanism for Bni1- and mDia1-induced actin assembly and dual regulation of Bni1 by Bud6 and profilin. Mol Biol Cell 15:896–907

Nakano K, Imai J, Arai R et al 2002 The small GTPase Rho3 and the diaphanous/formin For3 function in polarized cell growth in fission yeast. J Cell Sci 115:4629–4639

Pelham RJ, Chang F 2001 Role of actin polymerization and actin cables in actin-patch movement in *Schizosaccharomyces pombe*. Nat Cell Biol 3:235–244

Radcliffe P, Hirata D, Childs D, Vardy L, Toda T 1998 Identification of novel temperature-sensitive lethal alleles in essential beta-tubulin and nonessential alpha2-tubulin genes as fission yeast polarity mutants. Mol Biol Cell 9:1757–1771

Rodriguez OC, Schaefer AW, Mandato CA et al 2003 Conserved microtubule-actin interactions in cell movement and morphogenesis. Nat Cell Biol 5:599–609

Sawin KE, Nurse P 1998 Regulation of cell polarity by microtubules in fission yeast. J Cell Biol 142:457–471

Small JV, Geiger B, Kaverina I, Bershadsky A 2002 How do microtubules guide migrating cells? Nat Rev Mol Cell Biol 3:957–964

Snaith HA, Sawin KE 2003 Fission yeast mod5p regulates polarized growth through anchoring of tea1p at cell tips. Nature 423:647–651

Snell V, Nurse P 1993 Investigations into the control of cell form and polarity: the use of morphological mutants in fission yeast. Dev Suppl 289–299

Snell V, Nurse P 1994 Genetic analysis of cell morphogenesis in fission yeast—a role for casein kinase II in the establishment of polarized growth. EMBO J 13:2066–2074

Tanaka E, Kirschner MW 1995 The role of microtubules in growth cone turning at substrate boundaries. J Cell Biol 128:127–137
Toda T, Umesono K, Hirata A, Yanagida M 1983 Cold-sensitive nuclear division arrest mutants of the fission yeast *Schizosaccharomyces pombe*. J Mol Biol 168:251–270
Tran PT, Marsh L, Doye V, Inoue S, Chang F 2001 A mechanism for nuclear positioning in fission yeast based on microtubule pushing. J Cell Biol 153:397–411
Vardy L, Toda T 2000 The fission yeast gamma-tubulin complex is required in G(1) phase and is a component of the spindle assembly checkpoint. EMBO J 19:6098–6111
Verde F, Mata J, Nurse P 1995 Fission yeast cell morphogenesis: Identification of new genes and analysis of their role during the cell cycle. J Cell Biol 131:1529–1538
Wallar BJ, Alberts AS 2003 The formins: active scaffolds that remodel the cytoskeleton. Trends Cell Biol 13:435–446
West RR, Malmstrom T, Troxell CL, McIntosh JR 2001 Two related kinesins, klp5+ and klp6+, foster microtubule disassembly and are required for meiosis in fission yeast. Mol Biol Cell 12:3919–3932
Yaffe MP, Harata D, Verde F et al 1996 Microtubules mediate mitochondrial distribution in fission yeast. Proc Natl Acad Sci USA 93:11 664–11 668

DISCUSSION

Sheetz: You find those complexes at the side early on, in the rsp1 mutants. Why doesn't it activate when it is at the side of the cell early on?

Chang: The rsp1 cells, which have altered MT organization, often do start to grow strangely (Zimmerman et al 2004). They have a growth polarity defect in themselves. This is in part because the MTs are delivering these polarity factors to abnormal sites.

Kaibuchi: How does tea1 anchor at the cortex?

Chang: There is a protein called mod5, which is involved in docking tea1 at the cortex (Snaith & Sawin 2003). mod5 itself has a lipid anchor. It is restricted to a small region at the cell tips, but we don't know why. It probably has something to do with membrane domains and lipid rafts, which may be located at cell tips.

Alberts: Is for3 autoregulated?

Chang: What do you think?

Alberts: I suspect it is. So my next question is that do you think a GTPase is going to be sufficient to activate it? The evidence for GTPase activation of formins is still unclear.

Chang: At the moment we don't know. Your elegant model suggests that this may be one way of activating formins, but there certainly could be other mechanisms. What we would like to do is express these proteins *in vitro*, and see for example whether tea4 or a Rho-GTPase regulates the activity of this formin.

Alberts: That is what I like about your models. Why not have tea4 being the thing that is missing from the observation that you can't activate an autoregulated formin? Perhaps bud6 is at the other end and contributes to activation along with the Rho protein. Where is the GTPase in this context?

Chang: In general, several Rhos are located at the growing tips.

Alberts: The FH3 domain is supposed to be required for localization. Could this all be due to localization, and not necessarily activation?

Chang: Yes, this regulation by MTs could be primarily through localizing the formin to the right site. However, there is a hint that tea4p can also affect the activity of the formin. We have this intriguing observation that if we overexpress tea4 we are stimulating the formin massively. If we overexpress formin or tea1 we don't get this effect. It is not clear how tea4 is doing this (S. Martin and F. Chang, unpublished observations).

Alberts: Addressing the autoregulation problem, have you taken a truncated version of for3, and will that rescue?

Chang: We have tried overexpressing truncations of for3p and do not detect evidence for deregulated formin activity (S. Martin and F. Chang, unpublished observations).

Alberts: The same thing works for Bni1 in *Saccharomyces cerevisiae.*

Surana: Are cytokinesis sites misplaced in mutants defective in NETO?

Chang: No, the polarization mechanisms at the cell tip appear to be largely independent of the cytokinesis mechanisms. We think the nucleus determines where the cell division site is set up. However, in NETO mutants (such as *tea1* and *tea4* mutants) the division plane seems to be slightly skewed to one side of the nucleus. We are not sure why.

Nelson: When you make a chimera between tea1 and for3, if you take tea4 out of the equation can you still rescue.

Chang: Exactly. We can argue tea4 acts to bring tea1 and for3 together, and that this interaction may be sufficient to trigger this positive feedback.

Alberts: This raises a question. You nicely showed that tip1 is deposited against the MTs and you said all the initiation was dependent on the MTs. Could it be that suddenly tip1 is distributed evenly and this sets up the environment where a newly assembled complex gets moved?

Chang: I don't understand.

Nelson: I will reconfigure his question! Have you looked at a *tea2* mutant, and if so, what happens to mal3 and tip1 then?

Chang: Not directly, but other labs have been looking at these proteins. (Browning et al 2003, Busch & Brunner 2004, Busch et al 2004). I think mal3 looks normal in tea2 mutants. Tip1 is evenly distributed as dots along the MT, as is tea1.

Nelson: So then these plus-end binding proteins may be regulated by kinesin, rather than the plus end itself.

Chang: Yes. In the three fungal organisms, this appears to be the case. I don't know whether it applies to mammals.

Nelson: We have seen EB1 moving along MTs and not just at the MT plus end in mammalian cells.

Borisy: Does tea1 move along MTs throughout the cell cycle, or only after the G2 signal?

Chang: It moves all the time along cytoplasmic MTs.

Borisy: Including on the side which is not yet activated?

Chang: Yes.

Borisy: This means that it is not the kinesin that is regulating it, at least in the sense of transport. Tea1 is being transported constitutively to the new end. But then something happens after the G2 signal. So tea1 is on throughout the cell cycle, but it doesn't have any effect on recruiting the formin until after the signal. I am trying to understand what the trigger for NETO is. You have told us a lot of events. Is it the formin recruitment per se?

Chang: Prior to NETO, tea1p is continuously deposited on the cell tips by MTs, even at the non-growing cell tip. So the transport of tea1p is not the limiting factor. Then something happens at NETO to trigger polarity. There must be a cell cycle signal that activates the system. Part of this is the recruitment of more for3p to the cell tip, as there is reduced formin at the non-growing cell tip before NETO and then there is more after NETO. We believe that the activity of proteins such as tea1p or tea4p may change at NETO to recruit the formin, and setting off the positive feedback cycle for cell polarization.

Borisy: In contrast with the story in budding yeast, here there is no problem in establishing the axis. The cell always has an axis. It is elongated.

Chang: Even though fission yeast cells are rod shaped, there are still active mechanisms to designate sites of polarization.

D Lane: The axis is dependent on the MT.

Chang: Without the MTs, the cell will still grow as a sort of cylinder, but it is distorted. It starts to form strange-shaped cells.

Borisy: What I mean is that just through the normal process of fission the daughter cell begins with an axis. It is a self-propagating axis. You don't have to start it anew. In mating or budding an axis has to be created from a unique point each cell cycle.

Mostov: Mating aside, in yeast there is a point there from the previous bud. We don't know how this works molecularly, but it is always there.

Borisy: How do you know that? I asked Matthias Peter this earlier and he said he didn't know.

Mostov: There is the bud scar.

Borisy: The bud scar is on the parent. I don't know whether it is on the daughter, but the daughter also establishes a unique axis.

Peter: They come apart at some point. I don't know that this is the signal for polarization.

Mostov: The bud forms there.

Peter: There is a correlation and it is going there, but I don't know the mechanism for this.

Borisy: If there is a tight correlation, then you can argue in the same way as for fission yeast that there is a unique point and there is a mechanism for propagating that point and therefore propagating the axis.

D Lane: Let me try to explore this idea, because I find it interesting. You said that in mating, the bud scar is not the point of polarization. It is dictated by the gradient that is established by mutual secretion of the two pheromones. How do you know that? Do the yeasts sit still enough that you can be confident of this conclusion?

Peter: You can establish a gradient with a needle, and film the cells on a microscope slide while they orient.

Mostov: Do these mating yeasts establish a new axis towards the pheromone?

Chang: They use the existing growing point and curve around to the pheromone.

D Lane: Thinking about the G2 point, the obvious candidate has to be a Cdk cyclin complex that is post S. Is there any evidence of phosphorylation of any of these components?

Chang: With all these components there is evidence for some phosphorylation. It hasn't been well characterized enough for me to say even whether it is cell cycle regulated.

Balasubramanian: I'm reminded of this kinase Ssp1, which is required for NETO. It seems like a good candidate for G2 regulation. Is there any idea of how this works?

Chang: Not really. There is a bunch of kinases that are potentially involved: Ssp1, Pom1 and also Pac kinase, which has been implicated in phosphorylating tea1 (Kim et al 2003).

Nelson: Coming back to MTs, in budding yeast Kar9 is down-regulated by phosphorylation at one of the spindle pole bodies (Liakopoulos et al 2003). On the other, Kar9 is not phosphorylated and is directed along MTs to the growing bud. Is there anything like that with those types of proteins here?

Chang: Not much is known. In the *mal3* (EB1) mutant the MTs are very short (Beinhauer et al 1997). There is no obvious Kar9 homologue.

Balasubramanian: In the *tea1* mutant, in some of the daughters there is NETO but not in others. How do you explain that?

Chang: After the cells divide one of the daughters grows at the old end, and the other daughter cell grows at what used to be the septum. If we look at the mother of the two cells, there is a growing end which was one that never grew before. By default, it might choose to grow at the end that has recently divided — the 'new end.'

Balasubramanian: This goes back to Gary Borisy's earlier point that the axis is already there.

Chang: It is not clear that this is what happens. There are other mutants which show different patterns.

D Lane: In the *tea4* deletion, tea1 and the formin went to the opposite poles. Do you have an explanation for this?

Chang: We don't understand this yet. There is for3 on one cell tip which is growing, and tea1 is on the other one. In this case MTs are carrying tea1 to both cell tips equally. However, tea1 is sticking at one end but not the other (S. Martin and F. Chang, unpublished results).

D Lane: Where is mod5?

Chang: It is exclusively at one end, but its localization is dependent on tea1 (Snaith & Sawin 2003, S. Martin and F. Chang, unpublished results).

D Lane: Earlier on you implied that mod5 was pre-polarized. I got the impression that it was always at each end and you didn't understand that. But now you are saying that in this model that mod5 is dependent on tea1.

Chang: Tea1 and mod5 are codependent for localization (Snaith & Sawin 2003). Without tea1, mod5 is delocalized.

Drubin: If you make a pheromone gradient perpendicular to the long axis, what happens?

Chang: I think it always chooses one of the poles, but I haven't looked at this directly.

Sheetz: If you do a double mutation of *tea4* and *tea1*, does for3 polarize?

Chang: The double mutant looks exactly the same as the tea1 mutant, with for3 located at one end of the cell (S. Martin and F. Chang, unpublished).

Borisy: What is the phenotype of the kinesin *kip2* mutation?

Chang: The kip2 kinesin is tea2 in *S. pombe* (Browning et al 2000). It has the same phenotype as the tea1 mutant, and has difficulties in transporting tea1 to the MT plus ends.

Borisy: So the motor is essential. Why is that the case? In the pre-microscopic era people thought that proteins could diffuse. In smaller cells there isn't far to go. Free diffusion would take a few hundred milliseconds at most. Why do you need a motor to get tea1 to a tip?

Chang: In part, it is almost negative regulation. You don't want tea1 everywhere in the cell; you want to control carefully where you put it. The cell has devised a way of sequestering tea1p on the MTs, so that there isn't much free diffusional tea1.

Firtel: When it heads out, what is the mechanism by which it is held and stabilized at the far end? After it goes down the railway track it gets to the end of the cell. How does it stabilize it there? To ask Gary's question a different way, if you have some kind of sink—a protein that it is interacting with—that is already prelocalized, then why not use a diffusion model?

Chang: My summer project was to do FRAP (fluorescence recovery after photobleaching) on tea1-GFP (F. Chang and D. Khodjakov, unpublished observations). If we FRAP tea1 at the cell tip where there is a lot of tea1, it comes back slowly, taking about 10 min, one dot at a time. This is what we expected: MTs are the only way of tea1 getting there. Then we added MBC, a MT inhibitor, and I expected that no tea1 should get here. What happened is that tea1 still came back, perhaps even faster. It came back not one dot at a time, but as a haze. All of a sudden the whole tip started glowing. This implied that MTs might be a sink that inhibit free diffusion.

Borisy: This doesn't answer Rick Firtel's question. If you have a sink at one end, then you don't need a transport mechanism: diffusion would be sufficient. The experiment you described is consistent with that interpretation.

Firtel: If you are using a transport mechanism to get something to the tip, either it has to stick somehow or you are bringing with it a mechanism by which it is sticking to the other side. There is a need for a stabilization mechanism. The FRAP experiments indicate that the sink is already present.

Chang: The binding site is there, and so yes, diffusion seems to work almost as well as active transport. But remember that without MTs, cells grow in a curved or bent manner. Thus, you need the MTs to fine-tune the localization so that the cell grows straight.

Sheetz: You don't need an all-or-none type model here. If there is any selective advantage in moving tea1 actively, then very rapidly the other mechanism will be lost. They only need to divide 2% faster than the cells without that mechanism and within 50 cell cycles they will be the majority of the population.

Firtel: I have a question about the interpretation of Fred Chang's FRAP experiment. If there is a sink there, how did it get there? Maybe it got carried initially with the tea4. How do you initially form that site at the end?

Chang: This polarity establishment is what I have been trying to talk about. At established cell tips, all the tea1-interacting factors are there, and thus can actively recruit more tea1p. This is the sink, I guess.

Alberts: Could the sink be generated from the other end? Instead of the signals being delivered saying 'come here', the signal could be at the other end saying 'go away'. It could be like a nuclear export signal.

Chang: That is an interesting idea.

Firtel: Is there any way to get MTs to take a right turn and end up on a lateral membrane? If this is possible, is a sink now formed there? Is a tea1 complex formed at that site?

Chang: Yes, this doesn't usually happen, but you can get situations where tea1 complex is deposited on the sides of cells, where it appears to induce cell polarization at this abnormal site, which leads to formation of branches (Sawin & Nurse 1998).

References

Beinhauer JD, Hagan IM, Hegemann JH, Fleig U 1997 Mal3, the fission yeast homologue of the human APC-interacting protein EB-1 is required for microtubule integrity and the maintenance of cell form. J Cell Biol 139:717–728

Browning H, Hackney DD, Nurse P 2003 Targeted movement of cell end factors in fission yeast. Nat Cell Biol 5:812–818

Browning H, Hayles J, Mata J, Aveline L, Nurse P, McIntosh JR 2000 Tea2p is a kinesin-like protein required to generate polarized growth in fission yeast. J Cell Biol 151:15–28

Busch KE, Brunner D 2004 The microtubule plus end-tracking proteins mal3p and tip1p cooperate for cell-end targeting of interphase microtubules. Curr Biol 14:548–559

Busch KE, Hayles J, Nurse P, Brunner D 2004 Tea2p kinesin is involved in spatial microtubule organization by transporting tip1p on microtubules. Dev Cell 6:831–843

Kim H, Yang P, Catanuto P et al 2003 The kelch repeat protein, Tea1, is a potential substrate target of the p21-activated kinase, Shk1, in the fission yeast, Schizosaccharomyces pombe. J Biol Chem 278:30 074–30 082

Liakopoulos D, Kusch J, Grava S, Vogel J, Barral Y 2003 Asymmetric loading of Kar9 onto spindle poles and microtubules ensures proper spindle alignment. Cell 112:561–574

Matsusaka T, Hirata D, Yanagida M, Toda T 1995 A novel protein kinase gene ssp1+ is required for alteration of growth polarity and actin localization in fission yeast. EMBO J 14:3325–3338

Sawin KE, Nurse P 1998 Regulation of cell polarity by microtubules in fission yeast. J Cell Biol 142:457–471

Snaith HA, Sawin KE 2003 Fission yeast mod5p regulates polarized growth through anchoring of tea1p at cell tips. Nature 423:647–651

Zimmerman S, Tran PT, Daga RR, Niwa O, Chang F 2004 Rsp1p, a J domain protein required for disassembly and assembly of microtubule organizing centers during the fission yeast cell cycle. Dev Cell 6:497–509

Finding the way: directional sensing and cell polarization through Ras signalling

Atsuo T. Sasaki and Richard A. Firtel[1]

Section of Cell and Developmental Biology, Division of Biological Sciences and Center for Molecular Genetics, University of California, San Diego, 9500 Gilman Drive, La Jolla, 3801 CA 92093, USA

Abstract. Chemotactic eukaryotic cells have the unique ability to sense a shallow extracellular chemoattractant gradient and translate it into a steep intracellular gradient. For example, phosphoinositide-3,4,5-trisphosphate (PIP3), the product of phosphatidylinositol-3-kinase (PI3K), is accumulated at the leading edge but not the back of a polarized chemotaxing cell. This is partially controlled by the reciprocal, preferential localization of PI3K and PTEN to the membrane at the front and back, respectively. However, upstream events that control the localized activation and localization of PI3K and PTEN remain unclear. Recent findings indicate that Ras is important for activation of the PI3K pathway and regulation of directed cell movement and cell polarity. Ras is activated at the leading edge, and this local activation occurs without asymmetric localization of PI3K and PTEN or the F-actin cytoskeleton. In contrast, PI3K localization is driven by F-actin polymerization. Thus, Ras functions as an essential part of the cell's compass acting upstream of PI3K while reciprocal localization of PI3K and PTEN amplify the PIP3 gradient, rather than create it. These observations suggest a positive feedback loop to amplify an initial PIP3 gradient in which localized F-actin polymerization recruits cytosolic PI3K to the leading edge, where it is activated by Ras to locally produce PIP3 that induces F-actin polymerization.

2005 Signalling networks in cell shape and motility. Wiley, Chichester (Novartis Foundation Symposium 269) p 73–91

Chemotaxis, directional cell movement, is a fundamental cellular process and plays an essential role in development, tissue homeostasis, wound healing, innate immunity and metastasis of tumour cells. The migratory systems can be classified into: (i) amoeboid crawling systems driven by F-actin assembly-induced force; and (ii) adhesion receptor-mediated cell movement, driven by remodelling of the

[1]This paper was presented at the symposium by Richard A. Firtel to whom correspondence should be addressed.

extracellular matrix, such as adhesion receptor and integrin-mediated attachment. This review focuses on the amoeboid crawling system, which evolved more than a billion years ago and is remarkably conserved between *Dictyostelium* and human leukocytes (Ridley et al 2003, Parent 2004).

Cell migration is a complex process that requires the coordinated remodelling of the actin cytoskeleton and cell adhesion. In amoeboid chemotaxis, including that performed by *Dictyostelium* and many classes of leukocytes, chemoattractants function through the activation of G protein-coupled receptors (GPCRs) and their coupled heterotrimeric G proteins and downstream effectors (Chung et al 2001a, Parent 2004). Cells are able to respond to very shallow gradients of chemoattractants, amplifying this signal to create a steep intracellular gradient of signalling molecules that ultimately induce extension of the leading edge and protrusion of pseudopodia in synchronization with retraction of the back (Ridley et al 2003). Myosin II is assembled at the cell's posterior and lateral sides, where it provides cortical tension and controls the retraction of the cell's posterior. Cells lacking myosin II or proteins such as PAKa, which is required for myosin II assembly, cannot properly retract their posterior and exhibit chemotaxis defects (Chung et al 2001b, Devreotes & Janetopoulos 2003, Worthylake & Burridge 2003, Xu et al 2003, Parent 2004). This review will describe our present understanding of cell polarization and directional cell movement in response to chemoattractant gradients, focusing on the role of Ras in mediating these processes.

Directional sensing, cell movement and cell polarization

Cells can respond to chemoattractant gradients as shallow as a 2–5% difference between the anterior and posterior of the cell. Therefore, cells must have the ability to compare differences in receptor occupancy and convert this shallow extracellular gradient into a steep intracellular gradient of signalling components (Chung et al 2001a, Iijima et al 2002, Merlot & Firtel 2003). Experimental observations suggest that there must be a signalling module or group of components, called a 'cellular compass' or 'directional sensing machinery', that impinges on and spatially orients cell polarization in and movement up a chemoattractant gradient. Although the components of the cellular compass and their mechanism of action have yet to be elucidated, data accrued over the last five years suggest that the sensing machinery lies downstream of G protein activation and upstream of the generation of phosphoinositide-3,4,5-trisphosphate (PIP3). This rationale is based on the observations that chemokine receptors are uniformly distributed around the entire plasma membrane (Chung et al 2001a, Iijima et al 2002), whereas PIP3 is highly localized to the leading edge of cells (Funamoto et al 2001, Weiner et al 2002). Although the G protein $\beta\gamma$ subunit exhibits a very shallow anterior–posterior gradient that mirrors receptor occupancy, this gradient is too shallow to explain the steep generation of PIP3 (Fig. 1). Furthermore, fluorescence resonance energy transfer (FRET) analyses indicate that the G protein α and $\beta\gamma$ subunits remain dissociated as long as receptors are occupied by a chemoattractant (Janetopoulos et al 2001), whereas PIP3 generation and phosphatidylinositol-3-kinase (PI3K) activation are transient in response to global (uniform) stimulation, indicating that these pathways adapt even if receptor occupancy by the chemoattractant persists (Funamoto et al 2001, 2002, Janetopoulos et al 2001, Huang et al 2003). Recent studies have begun to uncover the mechanism of local amplification of PIP3.

PIP3, the key to directional cell movement and cell polarization

Numerous reports implicate PI3Ks as important mediators of chemotaxis and directional movement. Pharmacological inhibition of PI3K in *Dictyostelium* and a variety of mammalian cell types causes the inhibition of chemotaxis and cell migration. *Dictyostelium* cells lacking PI3K1 and PI3K2, neutrophils derived, from mice lacking PI3Kγ, and macrophages in which PI3Kδ has been ablated, have defects in cell polarity and reduced efficiency of chemotaxis and directionality (Chung et al 2001b, Funamoto et al 2001, Fruman & Cantley 2002). At the inner face of the plasma membrane, phosphatidylinositol-3,4-bisphosphate (PIP2) and PIP3 are generated by PI3Ks and degraded by the 3-phosphoinositide phosphatase PTEN. PIP2 and PIP3 diffuse 100 times slower than aqueous-soluble second messengers, such as cAMP, cGMP and inositol-3,4,5-trisphosphate (IP3), which helps PIP2 and PIP3 to be locally amassed to very high concentrations (Postma & Van Haastert 2001). When PTEN levels are depleted, PIP3 levels increase, indicating that PTEN plays a critical role in regulating PIP3 levels, qualitatively, temporally and spatially along the membrane (Funamoto et al 2002, Iijima & Devreotes 2002). The accumulation of PIP3 leads to the rapid localization and activation of PH domain-containing proteins, including Akt/PKB, CRAC, PhdA, canonical Rho guanine-nucleotide exchange factors (RhoGEFs), and molecules able to bind PIP3, such as WAVE/SCAR (Oikawa et al 2004). Herein, the steep PIP3 gradient is converted to local enzymatic events at the leading edge. Consistent with these observations, activated Rac localizes to the region where PIP3 accumulates (Weiner et al 2002, Ridley et al 2003, Oikawa et al 2004). These observations confirm the coupling of PI3K to the action of Rho GTPases. Other PI3K effectors, such as Akt/PKB and PhdA, also regulate chemotaxis. These molecules are recruited to the leading edge through their PH domains. Disruption of Akt/PKB or PhdA in *Dictyostelium* results in strong cell polarity and chemotaxis defects (Meili et al 1999). Akt/PKB regulates cell polarity at least partly through activation of PAKa, which is essential for myosin II assembly (Chung et al 2001b). PhdA has been suggested to act as an adaptor molecule that recruits other molecules to the plasma membrane and is required for full induction of F-actin polymerization (Funamoto et al 2001).

Antagonism between PI3K and PTEN is important but not essential for creating the PIP3 gradient

A question central to understanding chemotaxis is how cells preferentially accumulate PIP3 at the leading edge. The first insight into the mechanisms of the spatial localization of PIP3 has come from studies of the subcellular distribution of PI3Ks and PTEN in *Dictyostelium* (Funamoto et al 2002, Iijima & Devreotes 2002, Fig. 1). *Dictyostelium* PI3K transiently translocates to the plasma membrane in response to global chemoattractant stimulation and to the leading edge in chemotaxing cells. This PI3K translocation does not require its kinase activity, C2 domain or Ras binding domain, but requires a domain located in the first 100 residues at the N-terminus of the protein that lacks homology to known functional domains (Funamoto et al 2002, K. Takeda, R. A. Firtel, unpublished data). PTEN exhibits a spatial localization reciprocal to that of PI3K. In the basal state, PTEN is localized at the plasma membrane, but after chemoattractant stimulation, it transiently delocalizes from the plasma membrane and moves to the cytosol (Funamoto et al 2002, Iijima & Devreotes 2002). During chemotaxis, PTEN is excluded from the leading edge and becomes localized to the side and back of the cell. Loss of PTEN results in extended and non-spatially restricted PI3K activity. *pten* null cells exhibit expansion of PIP3 and F-actin accumulation at the membrane, demonstrating that PTEN is required for temporally and spatially restricting both PI3K activity and F-actin synthesis at the leading edge (Funamoto et al 2002, Iijima & Devreotes 2002). Subsequent studies in neutrophils have confirmed that recruitment of PI3K to and loss of PTEN from the leading edge are conserved in some amoeboid chemotaxing mammalian cell types (Wang et al 2002, Li et al 2003, Gomez-Mouton et al 2004).

These translocalization events create the strong asymmetry in PIP3 signalling that leads to directional movement. However, PI3K and PTEN are not essential components for directional sensing. *Dictyostelium* cells lacking PI3K1/2 or neutrophils lacking PI3Kγ have undetectable Akt/PKB activation and severely reduced cell polarization and are highly defective in directional movement but cells are still able to move toward the chemoattractant source. Furthermore, the N-terminus of PI3K still localizes to the leading edge in *pi3k1/2* null cells (Funamoto et al 2002). These observations suggest that an event upstream of PI3K localization and activation senses the chemoattractant gradient.

Studies show that PH domain-containing proteins and PTEN localize in the presence of latrunculin A, a toxin that inhibits F-actin polymerization by sequestering G-actin (Parent et al 1998, Janetopoulos et al 2004, Sasaki et al

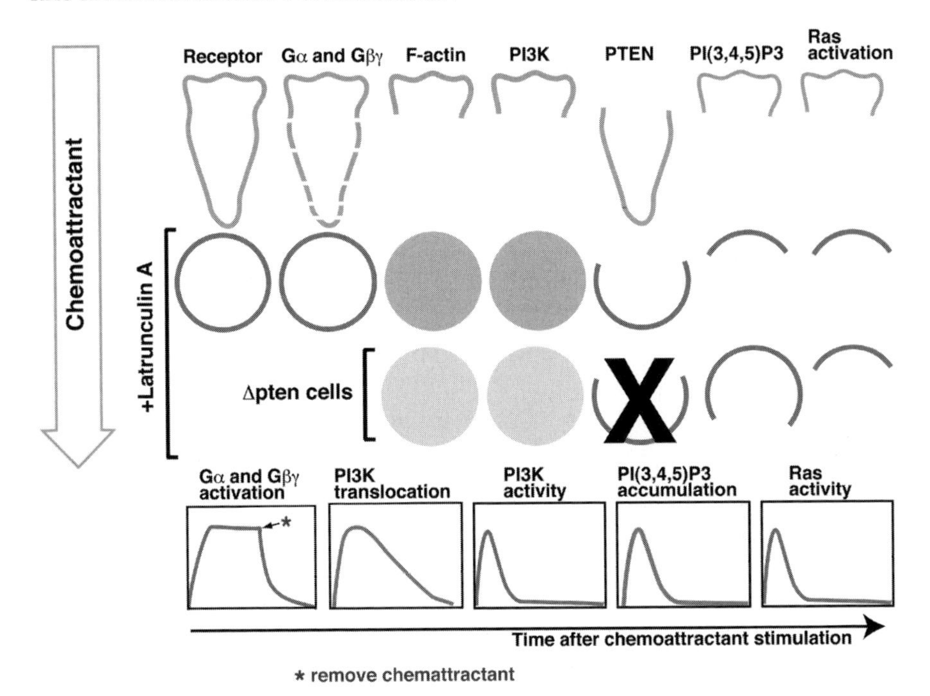

FIG. 1. Cellular distribution and activation profile of signalling components. The diagram shows the localization of signalling molecules in polarized wild-type (WT) chemotaxing cells, latrunculin A-treated wild-type cells, or *pten* null cells. The activation profile of each signalling molecule is shown at the bottom.

2004). Unexpectedly, the translocation of PI3K does not occur in latrunculin A-treated cells in GFP-based and cell fractionation assays (Sasaki et al 2004). Therefore, PI3K localization must require new F-actin polymerization. Thus, although PH domain proteins that act as reporters of PIP3 localize to the plasma membrane in response to chemoattractant stimulation in the absence of F-actin polymerization, the localization of PI3K, the enzyme responsible for PIP3 production, requires F-actin polymerization (Fig. 1). The explanations for these apparently contradictory observations are that a basal level of PI3K at the plasma membrane is activated upon chemoattractant stimulation and that the observed chemoattractant-dependent PI3K localization is part of a positive feedback loop (Sasaki et al 2004; see Discussion below).

Another key observation suggests the spatially restricted localization of PI3K and PTEN is not required for establishing the initial asymmetry of the chemoattractant response: localized PIP3 accumulation occurs in *pten* null cells treated with latrunculin A (Sasaki et al 2004). This shows that an intracellular PIP3 gradient or directional sensing event can be achieved without PI3K translocation and PTEN activity (Fig. 1). Although a link between F-actin-mediated PI3K localization and membrane rafts has been suggested (Gomez-Mouton et al 2004), the detailed mechanism underlying the localization of PI3K is still unknown. As PI3K localization partially overlaps with that of coronin, which binds newly synthesized actin, one plausible model is that PI3K preferentially binds to newly synthesized actin containing ATP.

Filling in the missing link between GPCR signalling and PIP3 production

In mammalian cells, heterotrimeric G protein-based signalling pathways predominantly activate the Class IB PI3K, PI3Kγ. Although Class IA PI3Ks are activated through p85- or p110-mediated membrane recruitment and interaction with Ras-GTP, activation of PI3Kγ is more complex (Fruman & Cantley 2002). In mammalian (HEK) cells, overexpression of G$\beta\gamma$ subunits leads to PI3Kγ activation by interacting with the p110γ catalytic subunit and the p110γ adaptor protein p101 (Brock et al 2003). G$\beta\gamma$ is expected to play an essential role in mediating the biochemical responses resulting in leading edge formation, but the absence of a highly polarized gradient of activated (free) G$\beta\gamma$ implies that it is not the signal responsible for intracellular amplification of the chemoattractant gradient and localized activation of PI3K. Furthermore, although the activation of the G protein is persistent, PI3K activation is transient (Janetopoulos et al 2001, Funamoto et al 2002, Huang et al 2003). These findings suggest that G$\beta\gamma$ activation is a prerequisite or priming event for the activation of PI3Kγ and indicate that a different component regulates the localized turning on and off of PI3K. Recent reports implicate Ras as an upstream regulator of PI3K.

Pathways from heterotrimeric G proteins to Ras

Heterotrimeric G proteins can engage multiple signalling pathways leading to diverse biological consequences. These G proteins also activate Ras via several different pathways: (i) epidermal growth factor (EGF)-shedding-mediated transactivation of EGFR, (ii) Src-mediated transactivation of receptor tyrosine kinases, (iii) Ca^{2+}-mediated RasGEF (RasGRFs and RasGRPs), and (iv) a novel pathway discussed later (Bivona & Philips 2003, Fig. 2). Ras has been proposed to be a general activator of mammalian and *Dictyostelium* PI3K. In mammalian cells, PI3Kγ is directly activated both in *vitro* and in *vivo* by Ras (Suire et al 2002), while in *Dictyostelium*, a functional Ras binding domain (RBD) is necessary for PI3K activation following chemoattractant stimulation (Funamoto et al 2002). Consistently, reports have shown that Akt/PKB activation is severely reduced in *rasGEF* knockout cells and cells expressing dominant negative Ras (Lim et al 2001, Sasaki et al 2004).

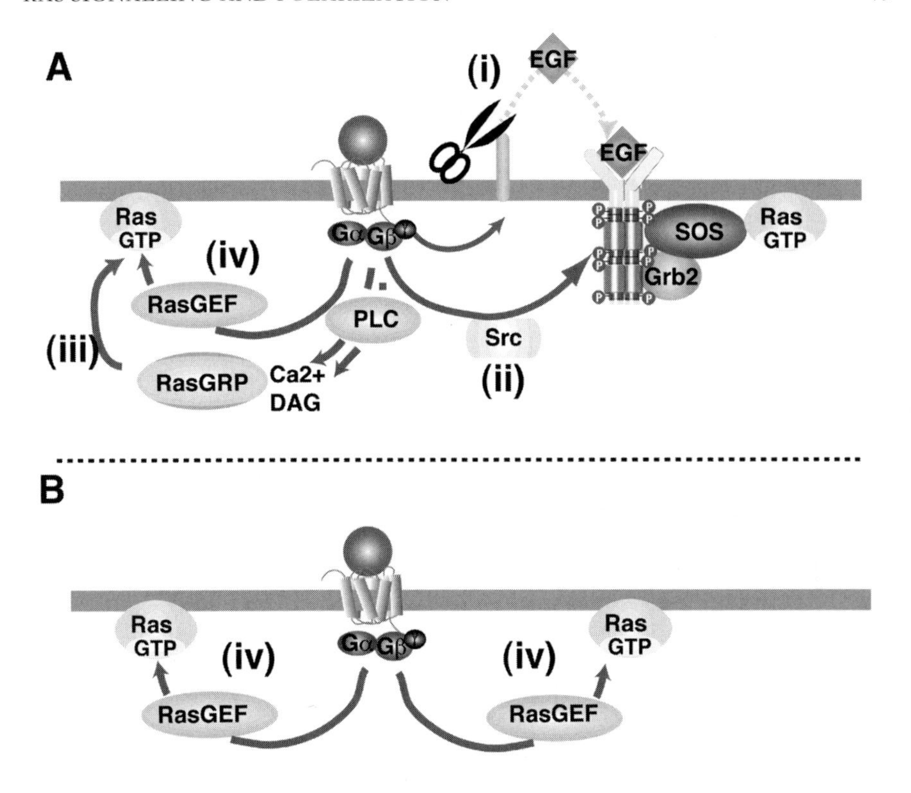

FIG. 2. Pathways for GPCR-Ras activation. (i) Transactivation model. GPCR signalling induces activation of metalloprotease, which generates HB-EGF (heparin binding EGF) through extracellular proteolytic cleavage of proHB-EGF. HB-EGF binds and activates an EGF receptor, and the Grb2-SOS complex recruits to activated EGFR where it activates Ras by GTP loading. (ii) GPCR-triggered activation of cytoplasmic tyrosine kinases, such as Src and Pyk, which phosphorylate receptor tyrosine kinases. (iii) RasGRP, a member of RasGEF, is activated by Ca^{2+} and DAG (diacylglycerol) produced through GPCR-mediated PLC (phospholipase C) activation. (iv) A proposed novel pathway to activate Ras from GPCR. RasGEFs such as Aimless are presumably activated through heterotrimeric G proteins.

Localized Ras activation regulates local PIP3 production

The mammalian Ras family comprises H-Ras, N-Ras and K-Ras, all of which are frequently mutated in human cancer and leukaemia. *Dictyostelium* has five Ras proteins related to mammalian H-Ras and K-Ras. Among these, RasB, RasD and RasG are most closely related to their mammalian counterparts and have a conserved effector domain sequence (Chubb & Insall 2001). Ras has been analysed intensively since the early 1980s, and these studies provided compelling evidence for Ras having a critical role in cell growth, differentiation and survival.

However, the importance of Ras in chemotaxis, cell polarization and directional sensing remained unclear, presumably due to a lack of proper tools and assay systems. Now, the impressive advances of *in vivo* imaging technology have been combined with the powerful genetic system of *Dictyostelium* to elucidate Ras signalling during chemotaxis.

A biochemical assay in which the GST-fused RBD from human Raf1 kinase or yeast Byr2 was used as an indicator for Ras activation demonstrated that Ras is rapidly and transiently activated upon global stimulation with a chemoattractant in *Dictyostelium* cells (Kae et al 2004, Sasaki et al 2004). The rapid kinetics of Ras activation are similar to those of PI3K activation (Huang et al 2003). Ras activation requires neither PI3K activation nor PTEN, demonstrating that Ras is upstream of PI3K. Importantly, Ras is not activated in Gβ or Gα2 mutant cells (Kae et al 2004), and its activation level is severely reduced in *aimless* null cells, one of the RasGEFs in *Dictyostelium* (Sasaki et al 2004). Because *Dictyostelium* does not have an EGF-shedding system, Src-like tyrosine kinases or Grb2-like adaptor molecules, and Aimless does not contain any known domains in addition to the RasGEF catalytic domain, these findings suggest the existence of a novel heterotrimeric G protein-meditated Ras activation pathway (Fig. 2).

Where is Ras activated in living mammalian cells?

Ras activation has been assayed in living mammalian cells using FRET and the GFP-Raf1 RBD. The chimeric probe Raich-Ras that utilizes the K-Ras C-terminus as a membrane-targeting anchor indicates that EGF stimulation activates Ras on the plasma membrane in COS-1 cells (Mochizuki et al 2001). Recently, H-Ras and N-Ras, but not K-Ras, were found to be activated at the plasma membrane, the Golgi and the endoplasmic reticulum intracellular membranes (Bivona & Philips 2003). Ras activation on endomembranes is delayed (peaking at 10–20 min post-stimulation) and sustained, whereas that on the plasma membrane is rapid (peaking in <5 min) and transient. Endomembrane Ras activation requires palmitoylation of Ras (which does not occur on K-Ras) and RasGRP1, a RasGEF activated by Ca^{2+} and diacylglycerol (DAG). Future analysis of Ras activation in chemotaxing mammalian cells should provide insight into its local activation in a polarized setting.

The use of *Dictyostelium* has revealed Ras activation in chemotaxing cells

In response to global chemoattractant stimulation, Ras activation, assayed via an RBD-GFP reporter, occurs rapidly and transiently on the plasma membrane (Sasaki et al 2004). This activation pattern is similar to that of K-Ras activation on the plasma membrane. Interestingly, no *Dictyostelium* Ras identified to date has a palmitoylation site similar to K-Ras. The initial Ras activation occurs as early as, or slightly faster than, PI3K translocation. PIP3 accumulation at the membrane occurs after both events. In chemoattractant gradients, activated Ras is restricted to the cell's leading edge, whereas total Ras protein is uniformly distributed along the plasma membrane (Sasaki et al 2004; Fig. 3). When a micropipette containing the chemoattractant was moved to the opposite side of the cell, the activated Ras signal was rapidly lost from the initial site and concomitantly re-accumulated at the new site of the membrane closest to the micropipette. Cells expressing myristoylation site tagged-PI3K and *pten* null cells have multiple pseudopodia, and activated Ras is found at the membrane region about to protrude (Sasaki et al 2004). These studies reveal that activated Ras localizes to the leading edge and to membrane domains that will form a leading edge. Strikingly, directed Ras activation occurs without PI3K translocation and PTEN activity, suggesting that Ras activation could create an intracellular gradient upstream of PIP3 gradient generation.

Ras regulates directed cell movement and cell polarity

Studies of mutations in several of the *Dictyostelium* Ras genes revealed that Ras regulates directional movement and cell polarity (Chubb & Insall 2001, Sasaki et al 2004). Examination of *aimless* (RasGEF) null cells demonstrated that Ras-mediated directional sensing is operated independently from Ras-mediated cell polarization. *aimless* null cells exhibit reduced directionality without a significant loss of cell polarity or speed (Chubb & Insall 2001, Sasaki et al 2004). We recently identified another *Dictyostelium* RasGEF mutant that shows complete loss of cell polarity without loss of cell directionality, suggesting that different Ras proteins and/or differential regulation of Ras proteins may control two distinct aspects of chemotaxis (A. T. Sasaki, R. A. Firtel, unpublished data). *pten* null cells also exhibit a defect in directional cell movement. In both *aimless* and *pten* null cells, Ras is occasionally activated at sites other than the leading edge, where cells begin to produce aberrant pseudopodia (Fig. 3), suggesting that incorrectly spatially restricted Ras activation causes the directional defect in those cells (Sasaki et al 2004, A. T. Sasaki, R. A. Firtel, unpublished data). There are >20 putative RasGEFs in the *Dictyostelium* genome. We speculate that other RasGEFs regulate directional sensing in addition to Aimless, since further depletion of Ras function through the expression of dominant-negative Ras in *aimless* null cells results in severe directionality and cell polarity defects (Sasaki et al 2004). Specific RasGEFs may also activate certain subsets of Ras to regulate them in a spatially or temporally restricted manner.

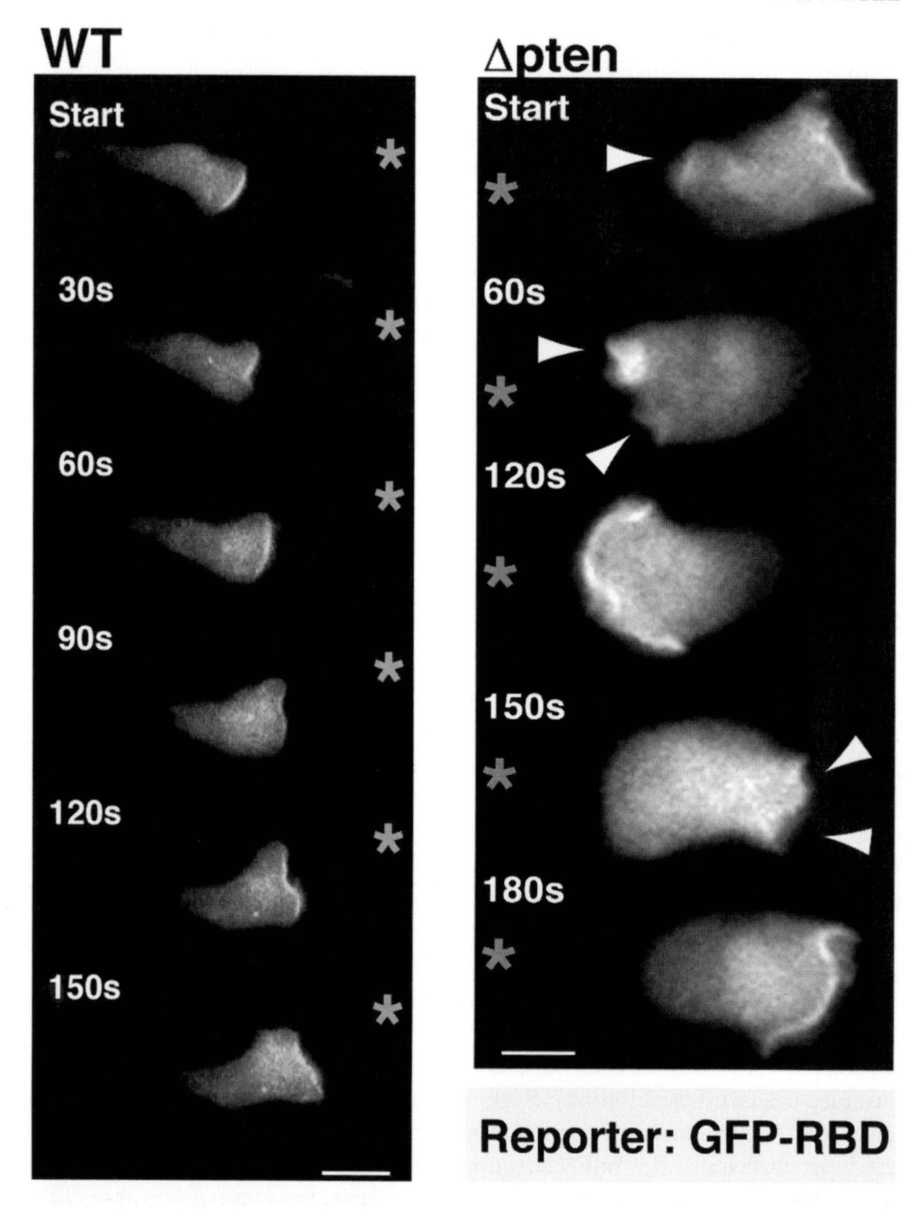

FIG. 3. Spatial–temporal activation of Ras during chemotaxis. The Ras activation in chemotaxing wild-type and *pten* null cells is imaged by GFP-RBD. An asterisk indicates the position of the micropipette. In *pten* null cells, occasionally, membrane localization of GFP-RBD was observed at the rear of a cell, and subsequently the cell started to move 'backward' relative to the chemoattractant gradient. The sites producing pseudopodia are marked with arrows.

Ras has multiple effectors in addition to PI3K, some of which control chemotaxis in a Ras-dependent manner. We identified RIP3 (Ras interacting protein 3) in *Dictyostelium* cells in a yeast two-hybrid screen using human Ras-GTP as bait, suggesting RIP3 is a Ras-interacting protein. *rip3* null cells exhibit a severe defect in cell polarization and reduced directional cell movement (Lee et al 1999). Unexpectedly, studies in yeast reveal that RIP3 is an orthologue of the yeast TOR complex 2 protein AVO1 (Loewith et al 2002). Mutation of *avo1* suppresses the effects of constitutively active Ras2, indicating that Ras is genetically coupled to the TOR complex 2 in yeast. The other TOR complex 2 proteins in *Dictyostelium* have been identified and their loss causes chemotaxis defects (S. Lee, R. A. Firtel, unpublished data). The phenotypes of *pi3k1/2* and *rip3* null cells suggest that proteins that bind Ras-GTP contribute to cell polarity through the PI3K and RIP3 pathways. We speculate that Ras regulates chemotaxis through PI3K, TOR and most likely other effectors.

The mechanism by which a shallow gradient is changed to a steep gradient

Several models have been proposed to explain how initial asymmetry in cells is generated and how cells accomplish directional sensing. While these models all include feedback loops, the feedback loops do not play a central role in the initial sensing of the gradient and establishing the initial asymmetry in the intracellular signalling pathways (Postma & Van Haastert 2001, Iijima et al 2002, Srinivasan et al 2003). As the finding of Ras/PI3K regulation has largely uncovered how the feedback loop functions, we now chose to review the feedback loop-mediated cell polarization and directional sensing (Fig. 4).

Signalling pathways are cascades of amplification in which the individual signalling molecules are both positively and negatively regulated. In each step of the pathway involving an enzyme, such as a GEF or kinase, the signal strength is greatly amplified. For chemotaxis, the initial amplification step is GPCR-mediated heterotrimeric G protein activation. However, this activation does not create a steep gradient and FRET analysis shows that only a very small portion of the cell's total G protein is activated (Janetopoulos et al 2001). Although it has not yet been analysed, we assume that single molecules of activated Gα2 or G$\beta\gamma$ might not activate very large numbers of downstream effectors, as G proteins basically bind to and activate effectors by inducing allosteric changes and/or by changing the localization of the effector. Therefore, RasGEF-mediated Ras activation could be the first enzymatic amplification step following GPCR activation, which is very steep and directed. The kinetics of Ras activation likely involve the activity of a Ras GAP(s), as Ras down-regulation also occurs rapidly.

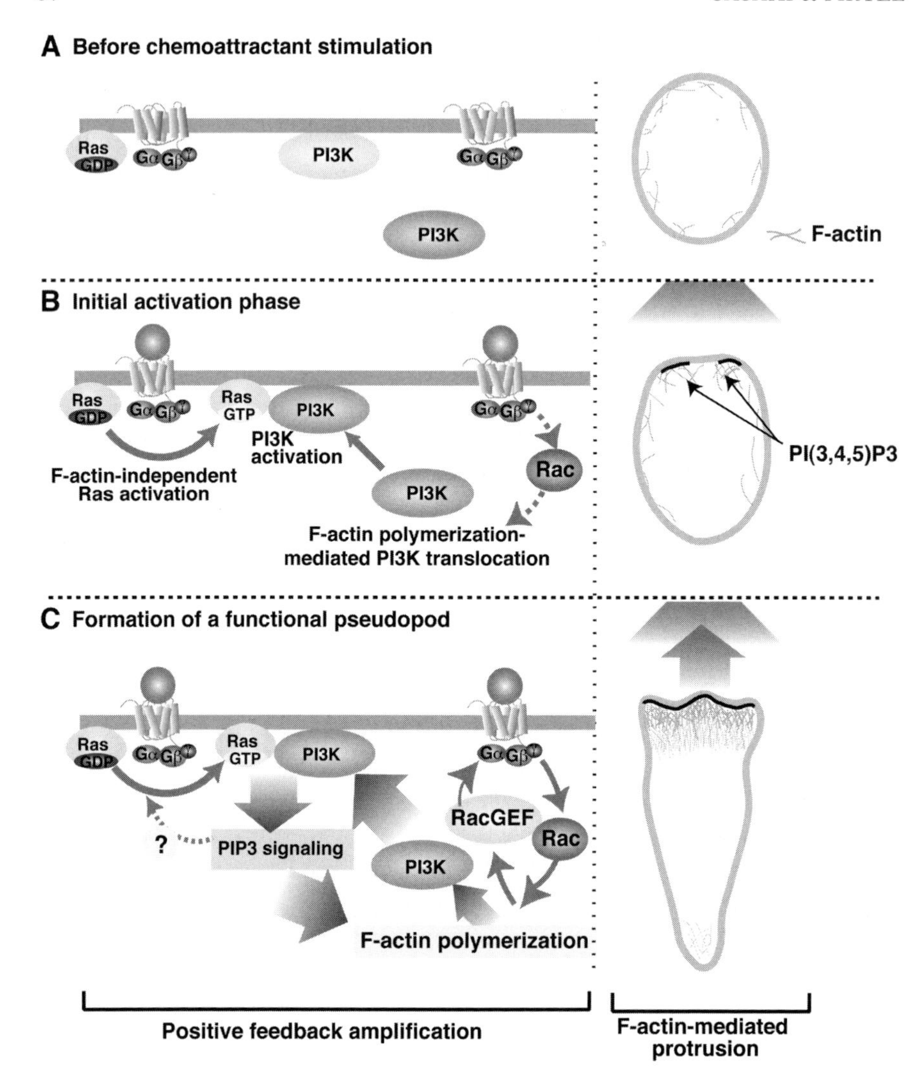

FIG. 4. A model for feedback loop-mediated directional sensing and cell polarization. The models show intracellular signalling to local PIP3 production (left), and chemoattractant-induced F-actin polymerization (right). The chemoattractant locally activates Ras at the presumptive leading edge (site of the membrane closest to the chemoattractant source) where Ras locally activates PI3K. There is a local polymerization of F-actin at the presumptive leading edge which is partially independent of the Ras/PI3K pathway, and we presume is controlled by Rho family members, WASP/SCAR and Arp2/3. F-actin mediates the translocation of PI3K. Locally produced PIP3 and F-actin-mediated RacGEF translocation induces further F-actin polymerization by activating downstream effectors, which enhances the recruitment of PI3K to the membrane.

PI3K activation would then be a second amplification event. Active PI3K produces PIP3 expansion, as long as Ras and $G\beta\gamma$ are activated.

A third amplification system consists of positive feedback loops between Ras activation and F-actin polymerization-induced PI3K translocation. It has been demonstrated that Ras activates PI3K and the PIP3 product of PI3K subsequently induces full F-actin polymerization by activating the Rac/WAVE/SCAR complex. F-actin assembly leads to further translocation of PI3K to the site (Wang et al 2002, Weiner et al 2002). A recent report shows that RacGEF translocates to the plasma membrane in an F-actin-dependent manner (Park et al 2004). This RacGEF localization could provide an additional positive feedback loop to amplify F-actin. We presume there are still numerous undiscovered positive feedback loops that amplify the signal, but experimental evidence suggests that the initial Ras/actin/PI3K/RacGEF amplification events cause cells to become polarized (Fig. 4). In addition, antagonistic mechanisms, such as the reciprocal localization of PI3K and PTEN, and reciprocal inhibition from the leading edge to the back and the lateral sides through Rac/F-actin and Rho/myosin II, play essential roles in establishing a stable leading edge and posterior. The Ras machinery is a key intermediate component between GPCR activation and strong PIP3 asymmetry, cell polarization, and directed cell movement. When a robust Ras activation meets F-actin synthesis, it is a moment to start making a new leading edge.

Acknowledgements

We gratefully acknowledge the members of Firtel laboratory for their stimulating discussions and helpful suggestions, Michelle Mendoza for critical comments, and Jennifer Roth for help in preparing this manuscript. ATS was supported, in part, by a Japanese Society for the Promotion of Science Research Fellowship for Research Abroad. This work was funded by research grants from the USPHS to RAF.

References

Bivona TG, Philips MR 2003 Ras pathway signaling on endomembranes. Curr Opin Cell Biol 15:136–142

Brock C, Schaefer M, Reusch H et al 2003 Roles of G beta gamma in membrane recruitment and activation of p110 gamma/p101 phosphoinositide 3-kinase gamma. J Cell Biol 160:89–99

Chubb JR, Insall RH 2001 Dictyostelium: an ideal organism for genetic dissection of Ras signalling networks. Biochim Biophys Acta 1525:262–271

Chung C, Funamoto S, Firtel R 2001a Signaling pathways controlling cell polarity and chemotaxis. Trends Biochem Sci 26:557–566

Chung C, Potikyan G, Firtel R 2001b Control of cell polarity and chemotaxis by Akt/PKB and PI3 kinase through the regulation of PAKa. Mol Cell 7:937–947

Devreotes P, Janetopoulos C 2003 Eukaryotic chemotaxis: distinctions between directional sensing and polarization. J Biol Chem 278:20 445–20 448

Fruman DA, Cantley LC 2002 Phosphoinositide 3-kinase in immunological systems. Semin Immunol 14:7–18

Funamoto S, Milan K, Meili R, Firtel R 2001 Role of phosphatidylinositol 3′ kinase and a downstream pleckstrin homology domain-containing protein in controlling chemotaxis in Dictyostelium. J Cell Biol 153:795–810

Funamoto S, Meili, R, Lee S, Parry L, Firtel R 2002 Spatial and temporal regulation of 3-phosphoinositides by PI 3-kinase and PTEN mediates chemotaxis. Cell 109:611–623

Gomez-Mouton C, Lacalle RA, Mira E et al 2004 Dynamic redistribution of raft domains as an organizing platform for signaling during cell chemotaxis. J Cell Biol 164:759–768

Huang Y, Iijima M, Parent C, Funamoto S, Firtel R, Devreotes P 2003 Receptor-mediated regulation of PI3Ks confines PI(3,4,5)P3 to the leading edge of chemotaxing cells. Mol Biol Cell 14:1913–1922

Iijima M, Devreotes P 2002 Tumor suppressor PTEN mediates sensing of chemoattractant gradients. Cell 109:599–610

Iijima M, Huang Y, Devreotes P 2002 Temporal and spatial regulation of chemotaxis. Dev Cell 3:469–478

Janetopoulos C, Jin T, Devreotes P 2001 Receptor-mediated activation of heterotrimeric G-proteins in living cells. Science 291:2408–2411

Janetopoulos C, Ma L, Devreotes PN, Iglesias PA 2004 Chemoattractant-induced phosphatidylinositol 3,4,5-trisphosphate accumulation is spatially amplified and adapts, independent of the actin cytoskeleton. Proc Natl Acad Sci USA 101:8951–8956

Kae H, Lim CJ, Spiegelman GB, Weeks G 2004 Chemoattractant-induced Ras activation during Dictyostelium aggregation. EMBO Rep 5:602–606

Lee S, Parent CA, Insall R, Firtel RA 1999 A novel Ras-interacting protein required for chemotaxis and cyclic adenosine monophosphate signal relay in Dictyostelium. Mol Biol Cell 10:2829–2845

Li Z, Hannigan M, Mo Z et al 2003 Directional sensing requires G beta gamma-mediated PAK1 and PIX alpha-dependent activation of Cdc42. Cell 114:215–227

Lim C, Spiegelman G, Weeks G 2001 RasC is required for optimal activation of adenylyl cyclase and Akt/PKB during aggregation. EMBO J 20:4490–4499

Loewith R, Jacinto E, Wullschleger S et al 2002 Two TOR complexes, only one of which is rapamycin sensitive, have distinct roles in cell growth control. Mol Cell 10:457–468

Meili R, Ellsworth C, Lee S, Reddy T, Ma H, Firtel R 1999 Chemoattractant-mediated transient activation and membrane localization of Akt/PKB is required for efficient chemotaxis to cAMP in Dictyostelium. EMBO J 18:2092–2105

Merlot S, Firtel R 2003 Leading the way: directional sensing through phosphatidylinositol 3-kinase and other signaling pathways. J Cell Sci 116:3471–3478

Mochizuki N, Yamashita S, Kurokawa K et al 2001 Spatio-temporal images of growth-factor-induced activation of Ras and Rap1. Nature 411:1065–1068

Oikawa T, Yamaguchi H, Itoh T et al 2004 PtdIns(3,4,5)P3 binding is necessary for WAVE2-induced formation of lamellipodia. Nat Cell Biol 6:420–426

Parent CA 2004 Making all the right moves: chemotaxis in neutrophils and Dictyostelium. Curr Opin Cell Biol 16:4–13

Parent C, Blacklock B, Froehlich W, Murphy D, Devreotes P 1998 G protein signaling events are activated at the leading edge of chemotactic cells. Cell 95:81–91

Park KC, Rivero F, Meili R, Lee S, Apone F, Firtel RA 2004 Rac regulation of chemotaxis and morphogenesis in *Dictyostelium*. EMBO J 23:4177–4189

Postma M, Van Haastert PJM 2001 A diffusion-translocation model for gradient sensing by chemotactic cells. Biophys J 81:1314–1323

Ridley AJ, Schwartz MA, Burridge K et al 2003 Cell migration: integrating signals from front to back. Science 302:1704–1709

Sasaki AT, Chun C, Takeda K, Firtel RA 2004 Localized Ras signaling at the leading edge regulates PI3K, cell polarity, and directional cell movement. J Cell Biol 167:505–518
Srinivasan S, Wang F, Glavas S et al 2003 Rac and Cdc42 play distinct roles in regulating PI(3,4,5)P3 and polarity during neutrophil chemotaxis. J Cell Biol 160:375–385
Suire S, Hawkins P, Stephens L 2002 Activation of phosphoinositide 3-kinase gamma by Ras. Curr Biol 12:1068–1075
Wang F, Herzmark P, Weiner O, Srinivasan S, Servant G, Bourne H 2002 Lipid products of PI(3)Ks maintain persistent cell polarity and directed motility in neutrophils. Nat Cell Biol 4:513–518
Weiner O, Neilsen P, Prestwich G, Kirschner M, Cantley L, Bourne H 2002 A PtdInsP(3)- and Rho GTPase-mediated positive feedback loop regulates neutrophil polarity. Nat Cell Biol 4:509–513
Worthylake RA, Burridge K 2003 RhoA and ROCK promote migration by limiting membrane protrusions. J Biol Chem 278:13 578–13 584
Xu J, Wang F, Van Keymeulen A et al 2003 Divergent signals and cytoskeletal assemblies regulate self-organizing polarity in neutrophils. Cell 114:201–214

DISCUSSION

Sheetz: What are the relative timetables for these various pathways? In other words, how rapidly is the activation occurring in this 2% gradient? The time for actin depolymerization and reorientation must be much longer than that initial activation.

Firtel: We have only done a bit of work looking at relatively shallow gradients. The experiments are much more difficult to do and the kinetics trickier to measure. We don't really know when the cells are actually sensing or seeing the signal. If we do the experiment in which we use reporters for PIP3 or PI3K itself, or the F-actin-binding protein coronin, which binds to new F-actin, we can quantitate what happens when we move the pipette from one side of the cell to the other. The loss of PH domain is very rapid, within a few seconds. If we use coronin and look at F-actin, this dissipates over about 4–8 s. The kinetics of appearance on the other side maximizes at about 10 s. The responses are still incredibly rapid.

Sheetz: That is with a high gradient.

Firtel: Yes, with a micropipette. The problem is that very few experiments have been done with shallow gradients. The new microfluid dynamic chambers will allow us to do this. The experiments are difficult, but I think we will be able to set up gradients of different concentrations. The initial suggestion is that the shallow gradient causes the cells to respond uniformly. Then they respond in the same way as the cringe response with actin polymerization: they adapt relatively rapidly. Then at the front of the cell, the side closest to the chemotactic source, we see a maintenance and accumulation of the PH domain reporter. This is where you get your pseudopod extending out. Part of the problem in thinking about these pathways is that the biochemistry is done with a global stimulation. You expose cells to chemoattractant and measure a biochemical response. The other responses are done with a directional signal. We try to take the data from one and the data from the other and put them together. We think we know from this how that pathway is working, but there are clearly limitations.

Titus: How do you think the F-actin polymerization is mediating the PIP3 translocation and stimulation?

Firtel: This has been a real problem. In neutrophils there is evidence that PREX, this PIP3-dependent exchange factor, may be a key regulator. In *Dictyostelium* it is known that the basic domain of WASP localizes in part to the membrane by PIP3 binding. If you make point mutations in that basic domain that abrogates the binding to PIP3, WASP will no longer localize. The other part of the pathway is that WASP needs to be activated and we need to get Rac activated. Even though there is a genetic dependence of Rac activation on PIP3, we don't know how this is regulated. It may be a secondary response because all of these amplification processes take place and PIP3 is required for this amplification process. The kinetics of Ras activation and PIP3 kinase localization are so fast that it is hard to separate them. We don't know, therefore, whether it is a primary or secondary response. We don't know how the initial Rac activation response is mediated. Is it $\beta\gamma$ (the most logical option) or is there another mechanism?

Borisy: I'd like to back-up for a moment. Your work shows something that was also discussed earlier, that there is a distinction between the directionality of a cell's response to an attractant and polarity. You described how the PI3K triple knockout would not polarize, but nevertheless it showed directionality towards a chemoattractant. Conversely, you described another mutant—the *aimless* null with the dominant-negative Ras—which could polarize but which didn't respond directionally. So you have experimentally uncoupled polarity from directional movement. Extending this to Meg Titus' question about the mechanism of feedback, the cells that are not polarizing in the absence of the PI3K, but which are responding directionally, must be activating actin polymerization on average towards the signal. So they are responding to this Rac activation.

Firtel: The argument we have is that this is three out of five PI3Ks. You still have a little bit there. You don't need much. But what you need the multiple PI3Ks for is the amplification of the localized signal at the front which gives you both polarity as well as directionality.

Borisy: We are still struggling to understand the reinforcement cycle. I would have thought that the amplification requirement would not have involved the cytoskeleton. It might have all been upstream of the cytoskeleton. But now you have introduced the cytoskeleton as part of the amplification process. Is that right?

Firtel: Yes. I would call it a two-stage amplification, though. If you look at the initial PIP3 response, which is PH domain localization, if you kill the cytoskeleton there is still an extremely localized PIP3 response. You don't need the cytoskeleton for that. We think this localized response is mediated through localized activation of Ras. If you have PI3K uniformly distributed and you locally activate Ras only in one spot, that is where you will get PIP3 production. We think the initial amplification is due to a biochemical response. There is a little bit of Ras activated (it is probably catalytic, but somewhat stoichiometric) but then you have PI3K cranking out PIP3. It keeps cranking this out, which will give an amplification. If you start with relatively few Ras molecules activating relatively few PI3K molecules, this then produces a large amount of PIP3 which binds a large number of PH domains. The actin cytoskeleton is not required for the initial directional response and the initial amplification of the external gradient, but it is required for subsequent amplification and the stabilization of the leading edge. We think this stabilization of the leading edge is what provides us with polarity and also persistence. By having all the components there, somewhat locked in with the cytoskeleton, permits persistence in movement.

Alberts: Can you make a distinction between needing F-actin intact versus new actin polymerization? Latrunculin will depolymerize actin which will block repolarization. Have you tried cytochalasin instead?

Firtel: I don't think we have a mechanism to do that. One of the things we tried to do is look at the *in vitro* binding of the localization domain of the PI3K and Rac. I asked some of my colleagues about how I could get ATP-bound F-actin. The most logical case in this type of response is that it is the ATP-F-actin that it might be binding to. This would be 'new' F-actin. People keep telling me that it can't be done in order to make it so that we can quantify the difference between binding and localization of stable F-actin versus new F-actin.

Peter: Your GEF doesn't have a PH domain, does it?

Firtel: Our GEF has the normal GEF PH domain. It has the VHPH double, but this PH domain doesn't bind PIP3, and it doesn't have an additional PH domain. It has a CH domain which is required for the localization. Technically we don't know it is binding to F-actin. We know that it is binding to the cortex and that the cortex requires F-actin.

Peter: Do you know for sure that it doesn't bind Ras?

Firtel: Binding to Ras doesn't have any impact on the localization. A Ras binding domain point mutation doesn't affect localization. If you remove the localization domain with the normal Ras binding domain the protein doesn't localize. Interestingly, if you try to use a Ras-binding domain for PI3K for these localization responses for Ras activation, you don't see them: the binding is too weak. My belief is that the activation response is not controlled through the localization of PI3K through activated Ras. It is a localization by a different mechanism. The Ras needs to interact and activate but I think it is a stabilized response rather than a localization response.

Harden: In neutrophils there is evidence that Rac/Cdc42 effector kinase PAK is being used as a scaffold for Cdc42 (Li et al 2003). Are there leading edge roles for PAK in *Dictyostelium*?

Firtel: There are four PAKs in *Dictyostelium*. PAKa, the one we have looked at the most, functions at the back to mediate myosin. PAKc functions at the leading edge, but the domains are not there and there is no evidence that it has a scaffolding function.

Harden: Do you have any RacGEFs that would bind to PAK?

Firtel: No. One of the PAKs is a myosin I heavy chain kinase. We believe that PAKc is a myosin I heavy chain kinase, because if we knock it out in the double mutant we see serious phenotypes, suggesting that they are possibly redundant on the same genetic pathway. The fourth PAK hasn't been looked at yet, but I don't think it has the domain structure to do what PAK1 is doing in neutrophils.

Harden: Why does *Dictyostelium* have so many Racs? How much functional information do we have on them?

Firtel: Dictyostelium lives by regulating its actin cytoskeleton, but it is a single cell. A single cell has to do everything that many of our cells do. It has to eat, move, undergo morphogenesis and so on. In order to control these processes separately it needs lots of GEFs and Racs. If you overexpress some of the Racs it will mediate the actin cytoskeleton randomly. If you knockout some of the other Racs the phenotypes don't affect chemotaxis. The cell is specialized.

Nelson: Have you tried to manipulate the levels of Ras in the context of how the cell senses small gradients of chemoattractants? Is this one of the key factors in whether the cell responds? Also, what about the GAPs? Is there a role for them in dampening down responses?

Firtel: The first question is more difficult to answer. We have been thinking of this as a threshold level. If you lower the Ras activity, you lower the activation response even though it has a relatively high threshold. This gives a non-polarized cell response. If you told one of the kids in your class that you were going to put a cell in a gradient with a 2% difference between the front and back of the cell, and asked them what the responses would be at each end, they would say that they were going to be identical. If you lose that polarity by affecting the threshold level, you can't get this amplification response and the polarization. You get a much more random activation around the cell. You lower the cytoskeleton response and get a random activation. We are now starting to look at the response to the GAPs. The basal level activation of constitutively active Ras is actually pretty low. In stimulated cells it goes way above wild-type, but the key issue is that it comes down much slower. We believe that the GAPs have to be a clear regulator. The same is true for the Rac proteins.

Alberts: Which version of activated Ras do you use?

Firtel: Predominantly Q61, but E12 gives a similar but weaker response.

Alberts: One might think that even constitutively active Ras molecules require exchange factors to bind nucleotide.

Firtel: That is what I am saying. They have a very low basal activation. A bit higher than wild-type, but not that much higher. The key thing is that it impacts the GAP component. They are much more stable in the GTP-bound form, and you get these phenotypes that are consistent with an extended activation pathway. If you look at PH domain in cells expressing constitutively active Ras, or AKT activation, it is extended and you get distinct sets of phenotypes.

Sheetz: The results from the delta PTEN mutant indicated that it was giving a non-polarized phenotype.

Firtel: In this null we think there are two aspects. First, we have much higher basal levels of PIP3 response. In chemotaxing polarized cells PTEN is missing from the front and is on the sides and back of cells. If we have any secondary response of activation on the sides of cells, it will be degraded by PTEN. Second, if we start out with a naïve cell with no PTEN there is a basal level of PIP3 activation that is uniform or random. When we activate cells we activate at much higher levels. Essentially, we believe the overall threshold of the response gets much lower so that a little bit of PI3K activation triggers a mass of uncontrolled response.

Reference

Li Z, Hannigan M, Mo Z et al 2003 Directional sensing requires G beta gamma-mediated PAK1 and PIX alpha-dependent activation of Cdc42. Cell 114:215–227

Roles of IQGAP1 in cell polarization and migration

Takashi Watanabe, Jun Noritake and Kozo Kaibuchi[1]

Department of Cell Pharmacology, Nagoya University, Graduate School of Medicine, 65 Tsurumai, Showa, Nagoya, Aichi, 466-8550, Japan

Abstract. Cell polarization and migration are fundamental processes in all organisms and are stringently regulated during tissue development, chemotaxis and wound healing. Migrating cells have a polarized morphology with an asymmetric distribution of signalling molecules and the cytoskeleton. Linkage of microtubule plus ends to the cortical region is essential for polarized migration. +TIPs, including CLIP-170 and APC (adenomatous polyposis coli) are thought to function as capturing devices at specialized cortical regions. Rho family GTPases, particularly Rac1 and Cdc42, play pivotal roles in cell polarization and migration acting through their effectors. We found that IQGAP1, an effector of Rac1 and Cdc42, interacts with CLIP-170. Activated Rac1 and Cdc42 enhance the binding of IQGAP1 to CLIP-170, and capture GFP-CLIP-170 at the base of leading edges and filopodia, respectively. Recently, we found that IQGAP1 directly binds to APC in addition to CLIP-170. IQGAP1 and APC interdependently localize to leading edges in migrating cells. IQGAP1 can link APC to actin filaments *in vitro*. Thus, activation of Rac1 and Cdc42 in response to migration signals leads to recruitment of IQGAP1 and APC which, together with CLIP-170, form a complex that links the actin cytoskeleton and microtubule dynamics during cell polarization and migration.

2005 Signalling networks in cell shape and motility. Wiley, Chichester (Novartis Foundation Symposium 269) p 92–105

Cell migration is necessary for developmental morphogenesis, tissue repair and regeneration, and tumour metastasis. Directional cell migration is usually initiated in response to extracellular cues such as chemokines and signals from the extracellular matrix. A migrating cell is highly polarized and contains complex regulatory pathways downstream of various stimuli. Cell polarization includes the asymmetric distribution of signalling molecules and cytoskeleton as well as directed membrane trafficking. A polarized migrating cell has a single

[1]This paper was presented at the symposium by Kozo Kaibuchi to whom correspondence should be addressed.

leading edge and filopodia at the front end; the microtubule-organizing centre (MTOC) and the Golgi apparatus are oriented toward the migrating direction, and temporal capture and stabilization of some microtubule plus ends occur near leading edges, which are thought to enable the cell to exert directed vesicular transport. Cell migration requires the coordination of these processes, especially the reorganization of the actin cytoskeleton and microtubules (Goode et al 2000, Rodriguez et al 2003).

Rho family GTPases play pivotal roles in cell polarization and migration through the regulation of the cytoskeleton and adhesion (Fukata et al 2003, Ridley et al 2003, Nelson 2003, Raftopoulou & Hall 2004). They function as molecular switches in cells by cycling between GDP-bound inactive and GTP-bound active forms to transmit various intracellular signals. The activity of Rho family GTPases is tightly regulated as described elsewhere (Takai et al 2001). Cdc42 and Rac1, members of Rho family GTPases, are implicated in cell polarization in the response to the various stimuli (Etienne-Manneville & Hall 2003a, Fukata et al 2003, Nelson 2003). Recently, the mechanism of actin cytoskeletal regulation by Cdc42 and Rac1 has been revealed through the identification and characterization of their effectors, such as N-WASP, PAK, IQGAP1 and myotonic-dystrophy-kinase-related Cdc42 binding kinase (Kaibuchi et al 1999).

Microtubules represent one of the major components of the cytoskeleton and are essential for cell division, cell migration, vesicle transport and cell polarization. Microtubules are nucleated from their minus ends, which localize predominantly at the MTOC. The plus ends of microtubules alternate between phases of growth and shrinkage (a state termed dynamic instability) to explore the intracellular spaces. Through this search process, the plus ends of microtubules are captured and stabilized at the target destinations, which include kinetochores on the mitotic spindle and cell cortex (Kirschner & Mitchison 1986, Schuyler & Pellman 2001a).

Recent studies have shown that some microtubule-associated proteins (MAPs) specifically accumulate at the plus ends of growing microtubules. These MAPs are termed +TIPs, which include CLIP-170, CLASPs, EB1, APC, dynein, dynactin and Lis1 (Schuyler & Pellman 2001b). The molecular mechanisms underlying the specific localization of +TIPs only at the plus ends of growing microtubules have been discussed elsewhere (Carvalho et al 2003). +TIPs are well conserved from yeast to mammals, and they play central roles in sensing cortical sites where a local signal activates microtubule-capping structures (Gundersen 2002). Genetic analyses in yeast revealed that these +TIPs are essential for the polarized microtubule array. Linkage of the plus ends of microtubules with the cell cortex is required for a polarized microtubule array that is characteristic of migrating cells, although it remains unclear how cortical molecules capture and stabilize the plus ends of microtubules during cell migration and whether Rho family GTPases, key regulators of cell polarization and migration, are involved in such capture and stabilization.

Cdc42 and Rac1 capture microtubules through IQGAP1 and CLIP-170

IQGAP1, an effector of Rac1 and Cdc42, is an integral protein of the actin cytoskeletal organization as well as cell migration (Briggs & Sacks 2003, Fukata et al 2003). We have shown that IQGAP1 localizes to basolateral membranes and regulates E-cadherin mediated cell–cell contacts both positively and negatively in epithelial cells (Fukata & Kaibuchi 2001, Noritake et al 2004). In migrating cells, IQGAP1 also accumulates at leading edges. IQGAP1 interacts with actin filaments through its calponin homology domain in the N-terminal region. Activated Rac1 and Cdc42 enhance the cross-linking activity of IQGAP1 for cell migration. However, the function of the C-terminal region of IQGAP1, which is conserved in its homologues, was not clarified. We identified CLIP-170 as an IQGAP1-interacting molecule by affinity column chromatography using mass spectral analysis (Fukata et al 2002). Immunoprecipitation assay showed that IQGAP1 forms a complex with CLIP-170 in Vero fibroblast cells. These results suggest that IQGAP1 interacts with CLIP-170 *in vivo*. We next performed an *in vitro* binding assay to examine whether IQGAP1 directly interacts with CLIP-170. We showed that the C-terminal region of IQGAP1 (1503–1657 aa) directly interacted with the N-terminal region (203–347 aa) of CLIP-170. As mentioned above, IQGAP1 accumulates at leading edges where actin meshwork is formed. We compared the localization of IQGAP1 and microtubules in Vero cells. IQGAP1 localized at leading edges where microtubules were targeted. IQGAP1 and microtubules were partially colocalized at leading edges, suggesting that IQGAP1 indirectly associates with microtubules through CLIP-170. Therefore, we examined whether IQGAP1 associates microtubules through CLIP-170 *in vitro*. IQGAP1, CLIP-170, and microtubules were mixed, and then microtubules were precipitated by centrifugation. We confirmed that CLIP-170 was coprecipitated with microtubules as previously described (Perez et al 1999). Under these conditions, IQGAP1 alone was not precipitated with microtubules. However, in the presence of CLIP-170, IQGAP1 was coprecipitated with microtubules. Thus, IQGAP1 appears to associate with microtubules through CLIP-170. Taken together with the results that IQGAP1 directly interacts with CLIP-170, this suggests that IQGAP1 can capture microtubules through CLIP-170 at a cortical region.

Do Rac1 and Cdc42 regulate the function of IQGAP1 and affect the complex formation among IQGAP1, CLIP-170 and microtubules? We showed that GTPγS-loaded Rac1 and Cdc42 increased the amount of associated IQGAP1

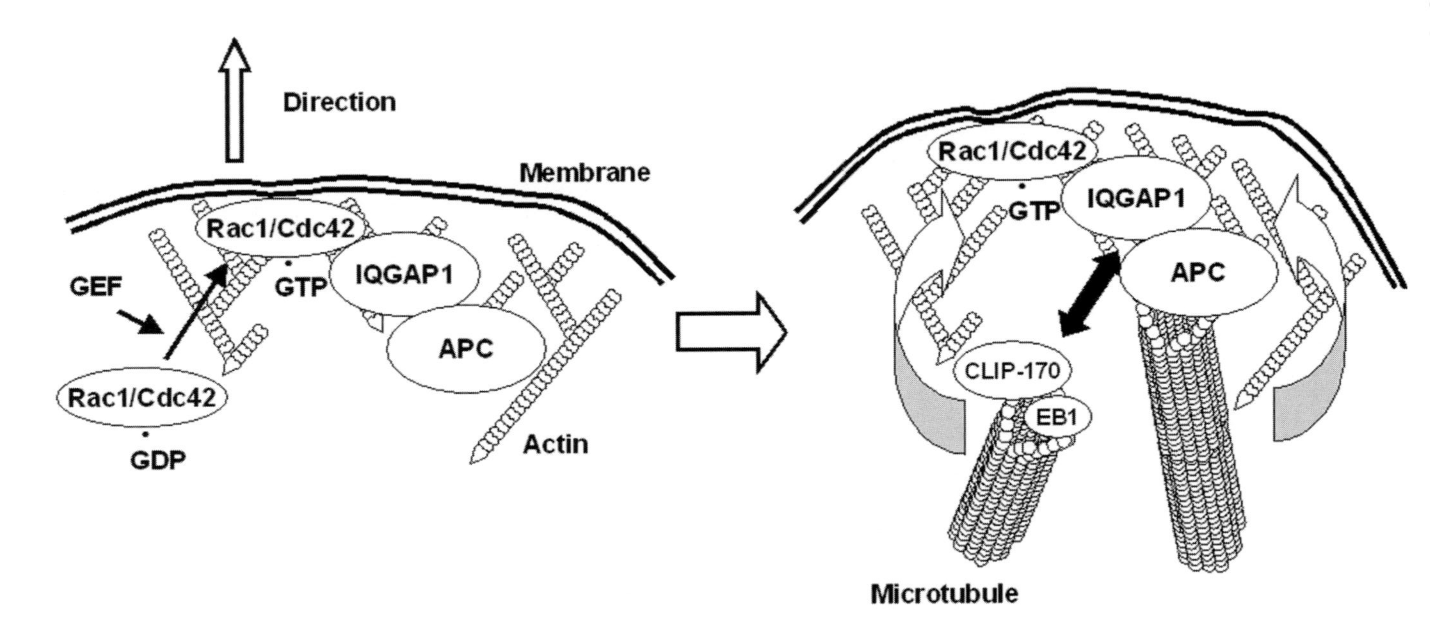

FIG. 1. Proposed model of the roles of IQGAP1 during cell polarization and migration. Activated Rac1 and Cdc42 induce the polymerization of actin filaments in concert with their effectors. Activated Rac1 and Cdc42 also determine the cortical spots where IQGAP1 cross-links actin filaments. There, APC is recruited through IQGAP1 to actin filaments. IQGAP1 captures the plus ends of microtubules via its interaction with CLIP-170. Then APC stabilizes microtubules directly and/or indirectly, which is necessary for the formation of stable actin meshwork at leading edges.

with microtubules through CLIP-170, in addition to forming a complex with IQGAP1 and CLIP-170. We next examined whether Cdc42 and Rac1 affect the dynamics of CLIP-170. Cells expressing EGFP-CLIP-170 were microinjected with constitutively active Rac1 or Cdc42. In control buffer-injected cells, most of EGFP-CLIP-170 disappeared as soon as it met a cortical region. When constitutively active Cdc42 was microinjected, filopodia were induced. At the base of Cdc42-induced filopodia, some EGFP-CLIP-170 remained immobile for about 2 min. When constitutively active Rac1 was microinjected, lamellipodia were induced. At the Rac1-induced lamellipodia, some EGFP-CLIP-170 remained immobile. These effects of Cdc42 and Rac1 were also observed at EGF (epidermal growth factor)- or HGF (hepatocyte growth factor)-induced leading edges. These results indicate that activated Rac1 and Cdc42 enhance the complex formation of IQGAP1 and CLIP-170, and thereby capture microtubules (Fig. 1).

Cdc42 has been shown to regulate the polarization of the MTOC by acting on Par6, aPKC (atypical protein kinase C), GSK-3β (glycogen synthase kinase 3β) and APC in migrating primary astrocytes during wound healing (Etienne-Manneville & Hall 2001, 2003b). The Par6-aPKC complex is an evolutionarily conserved cassette in cell polarity, and Cdc42 activates aPKC through Par6 (Etienne-Manneville & Hall 2003a, Macara 2004). APC and GSK-3β are involved not only in mediating the degradation of β-catenin under the control of Wnt signalling (Näthke 2004, Bienz 2002), but also in the stabilization of microtubules under the control of Cdc42. In fact, APC accumulates at the ends of microtubules that extend into actively migrating regions (Näthke et al 1996). APC also stabilizes microtubules, and phosphorylation by GSK-3β decreases APC-induced microtubule bundling (Zumbrunn et al 2001). Cdc42 activates aPKC via Par6, leading to phosphorylation and inactivation of GSK-3β at leading edges of migrating astrocytes, which suggest that inactivated GSK-3β allows APC to stabilize microtubules at leading edges. However, how and where APC is stabilized at the specialized cortical region remain to be clarified.

IQGAP1 links APC to cortical actin filaments during cell polarization and migration

Recently, we found that IQGAP1 forms a complex with APC *in vivo* (Watanabe et al 2004). *In vitro* binding assay showed that the C-terminal region of IQGAP1 directly binds to armadillo repeats of APC. In migrating fibroblasts, IQGAP1 and APC colocalized at leading edges where Rac1, Cdc42 and actin filaments accumulated. We examined whether activated Rac1 or Cdc42 forms a complex with IQGAP1 and APC. Immunoprecipitation and pull down assays showed that activated Rac1 and Cdc42 form a complex with IQGAP1 and APC. These results suggest that Rac1 and Cdc42 regulate a complex formation of IQGAP1 and APC.

To evaluate the function of IQGAP1 and APC in cell polarization and migration, we attempted RNAi (RNA interference) experiments using siRNA (small interfering RNA). During wound healing, control scramble-transfected cells showed a polarized morphology with a single leading edge where the actin meshwork was formed towards the wound. At this location, both IQGAP1 and APC were recruited and colocalized with actin filaments. The depletion of IQGAP1 impaired actin meshwork formation and decreased the accumulation of APC at leading edges. Similarly, the depletion of APC also prevented the accumulation of actin filaments and IQGAP1 at leading edges. The expression of IQGAP1 or APC fragment including the binding region of the counterpart also diminished the accumulation of APC or IQGAP1, respectively. Furthermore, the expression of $RNAi^R$ (RNAi resistant)-IQGAP1-WT or -APC-WT rescued the phenotypes created by each siRNA. These results suggest that IQGAP1 and APC interdependently localize to leading edges and IQGAP1 acts as an anchoring molecule for APC at the actin meshwork in the specialized cortical region.

Why is APC necessary for IQGAP1 localization at the leading edge? One explanation is that APC is needed for microtubule stabilization at the leading edge. Consistent with this, we showed that the treatment of cells with nocodazole, a microtubule-disrupting reagent, reduced the size of lamellipodia and decreased the accumulation of IQGAP1 and actin filaments there (Fukata et al 2002). We also found that $RNAi^R$-APC-WT rescued the inhibitory effects on the accumulation of IQGAP1/acitn filaments at the leading edge, whereas $RNAi^R$-truncated APC lacking the microtubule-binding domain did not. Thus, the binding of IQGAP1 and APC is direct, and the direct binding of APC to IQGAP1 is necessary for the accumulation of APC at actin filaments in the leading edge, whereas the effect of APC on IQGAP1/actin filaments can be mediated through the stabilization of microtubules.

Does the depletion of IQGAP1 and APC affect microtubules? We showed that activated Rac1 and Cdc42 immobilize EGFP-CLIP-170 at leading edges and the base of filopodia, respectively (Fukata et al 2002). We found that during wound healing most of EGFP-CLIP-170 was immobilized at leading edges toward the wound. The depletion of either IQGAP1 or APC impaired the immobilization of EGFP-CLIP-170 at leading edges. These results indicate that both IQGAP1 and APC are necessary for the stabilization of plus ends of microtubules.

In most migrating cells, microtubules are stabilized at leading edges. Dynein/dynactin is thought to pull the stabilized microtubules, to induce the MTOC reorientation towards the leading edge (Palazzo et al 2001a, Etienne-Manneville & Hall 2001). This reorientation facilitates directional cell migration. We also found that the depletion of either IQGAP1 or APC prevented the polarization of the MTOC. These inhibitory effects were not rescued by the expression of constitutively active Rac1 or Cdc42. The expression of the IQGAP1- or APC-binding fragment of APC or IQGAP1, respectively, impaired the reorientation of the MTOC. Furthermore, the expression of $RNAi^R$-IQGAP1-WT or $RNAi^R$-APC-WT rescued the reorientation of the MTOC, but that of $RNAi^R$ mutants which lack the binding region of the counterpart did not. These results indicate that IQGAP1 and APC are necessary for cell polarization and migration under the control of Rac1 and Cdc42.

We generated constitutively active IQGAP1, IQGAP1-T1050AX2, which has no binding activity to Rac1 and Cdc42 and a higher affinity to actin filaments. We showed that the expression of IQGAP1-T1050AX2 induced multiple leading edges where CLIP-170-decorated microtubules were targeted. As expected, IQGAP1-T1050AX2 was shown to have a higher affinity for CLIP-170 independent of Rac1 and Cdc42 (Fukata et al 2002). Recently, we also found that IQGAP1-T1050AX2 could provide the anomalous accumulation sites of APC. The expression of IQGAP1-T1050AX2 diminished the reorientation of the MTOC as well as directional cell migration during wound healing (Watanabe et al 2004). Thus, constitutively active IQGAP1 may recruit APC and CLIP-170 independently on Rac1 and Cdc42, leading to impaired cell polarization and migration through the additional formation of leading edges.

We showed that IQGAP1 directly interacts with CLIP-170 and APC in addition to actin filaments. Although it remained to be clarified how plus ends of microtubules are captured and how APC is anchored at the specialized cortical region, we proposed the model that IQGAP1 captures microtubules through CLIP-170 and IQGAP1 links APC to actin filaments during cell polarization and migration. Cell polarization and migration are usually initiated in the response to extracellular cues. Extracellular signals such as growth factors and chemokines activate Rac1 and Cdc42 through their guanine nucleotide exchange factors (GEFs) at leading edges. Activated Rac1 and Cdc42 induce the polymerization of actin filaments in the concert with their effectors, including WAVE and WASP. Activated Rac1 and Cdc42 also mark spots where IQGAP1 cross-links actin filaments. There, APC is recruited through IQGAP1 to actin filaments. IQGAP1 captures the plus ends of microtubules through CLIP-170 in concert with APC. Then, APC stabilizes microtubules directly and/or indirectly, which is necessary for a stable actin meshwork at leading edges (Fig. 1).

Stabilization of microtubules by Rho family GTPases

RhoA, another member of Rho family GTPases, induces the formation of Glu-tubulin which is a post-translationally modified tubulin that accumulates in stable microtubules (termed Glu-microtubules) with a half-life of 1 hour. RhoA, but not Rac1 and Cdc42, is necessary and sufficient for lysophosphatidic acid

(LPA)-induced Glu-microtubule formation through mDia, an effector of RhoA (Cook et al 1998, Palazzo et al 2001b). Furthermore, because mDia1 and mDia2 interact with APC and EB1, an APC-interacting +TIP, and mDia1 localizes at the tips of Glu-microtubules together with these proteins, the formation of Glu-microtubules can be regulated in the pathway of RhoA-mDia-APC-EB1-microtubule (Wen et al 2004). Activated Rac1 and Cdc42 capture CLIP-170 for a relatively short period at the cortical regions ($\sim$2 min) (Fukata et al 2002). The immobilization of CLIP-170 was observed in most microtubules that target to the periphery towards the wound (Watanabe et al 2004). Activated Rac1 also induces the stabilized microtubules partially through PAK and the inactivation of stathmin at leading edges in most cells (Daub et al 2001, Wittmann et al 2003, 2004). The Glu-microtubules, which orient to the wound, have a longer half-life of 1 hour and only 10% of microtubules in a cell show the accumulation of Glu-tubulin (Gundersen 2002). Although RhoA-induced stabilization is a consequence of the accumulation of post-translational modified tubulin, such modifications are not required for the capture and stabilization of microtubules by Rac1 or Cdc42. Rac1 and Cdc42 may affect the dynamics of microtubules in response to the signals which induce the reorganization of cytoskeleton. These regulatory pathways may regulate microtubules separately and have different physiological importance. More detailed analyses will be required to address these possibilities. In this regard, the activation of RhoA (the formation of Glu-microtubules) is dispensable, but Cdc42 signalling is necessary, for the reorientation of the MTOC toward the various stimuli (Palazzo et al 2001a, Etienne-Manneville & Hall 2001, Tzima et al 2003). Dynein and dynactin, in addition to the effectors of Cdc42, play essential roles in that process (Palazzo et al 2001a, Etienne-Manneville & Hall 2001). Dynein and dynactin may pull microtubules which are captured and stabilized by the activation of Cdc42. The mechanism by which the MTOC orients toward the direction of the stimuli needs to be characterized in more detail. Because +TIPs other than CLIP-170 and EB1 also accumulate at the plus ends of microtubules, it seems important to identify the +TIP-interacting molecules and investigate their association with the Rho family and their effectors. Although APC plays pivotal roles in the stabilization of microtubules, the detailed modes of the stabilization by APC are largely unknown. The stabilization of microtubules by APC is under the control of Cdc42, Rac1 and RhoA, but how these Rho family GTPases control the functions of APC and the specific stabilization effects of this +TIP on microtubules needs to be clarified.

Acknowledgements

We thank Dr F. Perez (Curie Institute) for providing CLIP-170 cDNA, Dr E. Mekada (Osaka University) for providing Vero cells, Dr T. Akiyama and Dr Y. Kawasaki (Tokyo University) for providing APC cDNA, Dr I. S. Näthke (University of Dundee) for providing anti-APC antibody. We also thank Dr T. Hakoshima (Nara Institute of Science and Technology) and Dr A. Kikuchi (Hiroshima University), and Dr H. Saya (Kumamoto University) for helpful discussion and preparing some materials. These researches were supported in part by Grants-in-Aid for Scientific Research, a Grant-in-Aid for Creative Scientific Research, and The 21st Century Centre of Excellence (COE) Program from MEXT, SCFPST, PMDA and The Nitto Foundation.

References

Bienz M 2002 The subcellular destinations of APC proteins. Nat Rev Mol Cell Biol 3:328–338

Briggs MW, Sacks DB 2003 IQGAP proteins are integral components of cytoskeletal regulation. EMBO Rep 4:571–574

Carvalho P, Tirnauer JS, Pellman D 2003 Surfing on microtubule ends. Trends Cell Biol 13:229–237

Cook TA, Nagasaki T, Gundersen GG 1998 Rho guanosine triphosphatase mediates the selective stabilization of microtubules induced by lysophosphatidic acid. J Cell Biol 141:175–185

Daub H, Gevaert K, Vandekerckhove J, Sobel A, Hall A 2001 Rac/Cdc42 and p65PAK regulate the microtubule-destabilizing protein stathmin through phosphorylation at serine 16. J Biol Chem 276:1677–1680

Etienne-Manneville S, Hall A 2001 Integrin-mediated activation of Cdc42 controls cell polarity in migrating astrocytes through PKCzeta. Cell 106:489–498

Etienne-Manneville S, Hall A 2003a Cell polarity: Par6, aPKC and cytoskeletal crosstalk. Curr Opin Cell Biol 15:67–72

Etienne-Manneville S, Hall A 2003b Cdc42 regulates GSK-3beta and adenomatous polyposis coli to control cell polarity. Nature 421:753–756

Fukata M, Kaibuchi K 2001 Rho-family GTPases in cadherin-mediated cell-cell adhesion. Nat Rev Mol Cell Biol 2:887–897

Fukata M, Watanabe T, Noritake J et al 2002 Rac1 and Cdc42 capture microtubules through IQGAP1 and CLIP-170. Cell 109:873–885

Fukata M, Nakagawa M, Kaibuchi K 2003 Roles of Rho-family GTPases in cell polarisation and directional migration. Curr Opin Cell Biol 15:590–597

Goode BL, Drubin DG, Barnes G 2000 Functional cooperation between the microtubule and actin cytoskeletons. Curr Opin Cell Biol 12:63–71

Gundersen GG 2002 Evolutionary conservation of microtubule-capture mechanisms. Nat Rev Mol Cell Biol 3:296–304

Kaibuchi K, Kuroda S, Amano M 1999 Regulation of the cytoskeleton and cell adhesion by the Rho family GTPases in mammalian cells. Annu Rev Biochem 68:459–486

Kirschner M, Mitchison T 1986 Beyond self-assembly: from microtubules to morphogenesis. Cell 45:329–342

Macara IG 2004 Par proteins: partners in polarization. Curr Biol 14:R160–162

Näthke IS 2004 The adenomatous polyposis coli protein: the Achilles heel of the gut epithelium. Annu Rev Cell Dev Biol 20:337–366

Näthke IS, Adams CL, Polakis P, Sellin JH, Nelson WJ 1996 The adenomatous polyposis coli tumor suppressor protein localizes to plasma membrane sites involved in active cell migration. J Cell Biol 134:165–179

Nelson WJ 2003 Adaptation of core mechanisms to generate cell polarity. Nature 422:766–774

Noritake J, Fukata M, Sato K et al 2004 Positive role of IQGAP1, an effector of Rac1, in actin-meshwork formation at sites of cell-cell contact. Mol Biol Cell 15:1065–1076

Palazzo AF, Joseph HL, Chen YJ et al 2001a Cdc42, dynein, and dynactin regulate MTOC reorientation independent of Rho-regulated microtubule stabilization. Curr Biol 11:1536–1541

Palazzo AF, Cook TA, Alberts AS, Gundersen GG 2001b mDia mediates Rho-regulated formation and orientation of stable microtubules. Nat Cell Biol 3:723–729

Perez F, Diamantopoulos GS, Stalder R, Kreis TE 1999 CLIP-170 highlights growing microtubule ends *in vivo*. Cell 96:517–527

Raftopoulou M, Hall A 2004 Cell migration: Rho GTPases lead the way. Dev Biol 265:23–32

Ridley AJ, Schwartz MA, Burridge K et al 2003 Cell migration: integrating signals from front to back. Science 302:1704–1709

Rodriguez OC, Schaefer AW, Mandato CA et al 2003 Conserved microtubule-actin interactions in cell movement and morphogenesis. Nat Cell Biol 5:599–609

Schuyler SC, Pellman D 2001a Search, capture and signal: games microtubules and centrosomes play. J Cell Sci 114:247–255

Schuyler SC, Pellman D 2001b Microtubule "plus-end-tracking proteins": the end is just the beginning. Cell 105:421–424

Takai Y, Sasaki T, Matozaki T 2001 Small GTP-binding proteins. Physiol Rev 81:153–208

Tzima E, Kiosses WB, del Pozo MA, Schwartz MA 2003 Localized cdc42 activation, detected using a novel assay, mediates microtubule organizing center positioning in endothelial cells in response to fluid shear stress. J Biol Chem 278:31 020–31 023

Watanabe T, Wang S, Noritake J et al 2004 Interaction with IQGAP1 links APC to Rac1, Cdc42, and actin filaments during cell polarization and migration. Dev Cell 7:871–883

Wen Y, Eng CH, Schmoranzer J et al 2004 EB1 and APC bind to mDia to stabilize microtubules downstream of Rho and promote cell migration. Nat Cell Biol 6:820–830

Wittmann T, Bokoch GM, Waterman-Storer CM 2003 Regulation of leading edge microtubule and actin dynamics downstream of Rac1. J Cell Biol 161:845–851

Wittmann T, Bokoch GM, Waterman-Storer CM 2004 Regulation of microtubule destabilizing activity of Op18/stathmin downstream of Rac1. J Biol Chem 279:6196–203

Zumbrunn J, Kinoshita K, Hyman AA, Näthke IS 2001 Binding of the adenomatous polyposis coli protein to microtubules increases microtubule stability and is regulated by GSK3 beta phosphorylation. Curr Biol 11:44–49

DISCUSSION

Nelson: When you express the C-terminal IQGAP to compete with CLIP-170, does this affect CLIP-170 loading onto microtubules and movement cortically, or only the cortical event?

Kaibuchi: If we express the C-terminal region of IQGAP1, CLIP-170 disappears everywhere in the cytoplasm.

Nelson: So you don't see it on microtubules at all?

Kaibuchi: No, we do not see the accumulation of CLIP-170 at the plus ends of microtubules.

Braga: Did you look at GSK-3β?

Kaibuchi: Not here. We are looking at the GSK-3β effect in neurons. We are just looking now at the effects of some GSK-3β inhibitors. They appear to stablilize CLIP-170 at the periphery. We assume that GSK-3β should be involved in this stabilization.

D Lane: Is it correct that there are two IQGAP isoforms?

Kaibuchi: No, there are actually three IQGAP isoforms.

D Lane: I seem to remember that IQGAP knockout mice run around quite happily.

Kaibuchi: Yes, they are almost normal, except for hyperplasia in their stomachs. IQGAP3 is widely expressed, whereas IQGAP2 is expressed exclusively in liver.

D Lane: What about in your cells?

Kaibuchi: They have a very small amount of IQGAP3.

D Lane: Physiologically, most cells will have one or more isoforms.

Kaibuchi: They have IQGAP1 and 3.

Borisy: I would like you to put your beautiful molecular experiments into a structural context. You say that IQGAP cross-links actin filaments. At the leading edge we think that actin filaments are organized in this dendritic network. In what sense is IQGAP cross-linking filaments? How do you relate the cross-linking you have suggested with the dendritic network?

Kaibuchi: We don't have enough data to be sure of this conclusion. We have previously shown that IQGAP1 cross-links F-actin in a cell-free system in the presence of activated Rac or Cdc42. So, we just imagine that IQGAP1 has the ability to cross-link F-actin in intact cells. When we knock down IQGAP1, GFP-actin accumulation is reduced. We think that IQGAP1 may be involved in cross-linking of F-actin.

Borisy: Cross-linking what to what? There is a structure present at the leading edge.

Kaibuchi: We need to see IQGAP1 present at the leading edge by electron microscopy and to show the relation between IQGAP1 and actin network structures, but we haven't done this yet.

Borisy: The second point I want to make is in connection with the recruitment of microtubules. There is ample evidence that the leading edge can advance without participation of microtubules. Can you tell us what you think the function of this recruitment is?

Kaibuchi: My assumption is simple. I think that actin filaments are essential for the leading edge formation, but microtubules are not essential. However, microtubules are important for maintaining the leading edges, because the leading edge is still induced if we treat the cell with nocodazole to eliminate microtubules, but the size of the leading edge and the accumulation of F-actin is reduced. Thus, microtubules are necessary to maintain the actin network at the leading edge. IQGAP1 recruits APC at the leading edges, and then APC stabilizes microtubules there. This may support the maintenance of the actin filaments, but we don't know the exact mechanism.

Sheetz: In terms of defining what you mean by 'leading edge', often there is a lamellipodium which is about 1.5 μm wide. This doesn't seem to contain microtubules. Back from that we often find the precursors of focal adhesions and so forth. Where do you fit this complex into that sort of architecture? Is it linking to those preliminary focal adhesions?

Kaibuchi: If you take a look at the leading edge carefully, there is a small space between actin filaments and microtubules. We usually observe the wide and thin lamellipodium. There, a small population of microtubules transiently appears to be connected to actin filaments. I believe that this connection is mediated by IQGAP1. We are currently working on the relationship between IQGAP1 and focal contacts and adhesions.

Firtel: The Rac is on the membrane.

Kaibuchi: Yes. We think that the immunoreactivity of Rac is on the membranes.

Vallee: CLIP-170 and the other microtubule tip-binding proteins are being accorded remarkable properties. We understand that CLIP-170 binds to the growing ends of microtubules. This is a transient phenomenon. If a microtubule stops in the middle of the cytoplasm CLIP-170, p150 and so on disappear fairly quickly. 'Fairly quickly' is an important issue because now you are finding that CLIP-170 persists for up to 2 minutes if it runs all the way to the cortex of the cell. I don't understand this completely. I can see that CLIP-170 would be stabilized itself at that site by binding to IQGAP, but it is still associated with the tip of the microtubule. This implies that its lifetime on the microtubule end is different from when the microtubule has hit the cortex.

Kaibuchi: Judging from the dynamics of CLIP-170, without IQGAP and APC, CLIP-170 disappears within 10 seconds. CLIP-170 is stabilized for up to 2 min at the leading edges in the presence of IQGAP1 and APC, whereas CLIP-170 is not stabilized at the opposite sites of the leading edges. Thus, we think that IQGAP1 and APC cooperatively stabilize CLIP-170 at the leading edges downstream of Rac1 and Cdc42.

Vallee: The point is, when you do have the interacting proteins, then I'm sure you said that CLIP-170 persists much longer than 10 s. This implies that its interaction with the microtubule is increased versus the free microtubule in the cytoplasm. This is what I want to clarify.

Kaibuchi: To clarify this we need to do double staining of microtubules and CLIP-170. We carried out experiments using cells expressing both DsRed-CLIP-170 and GFP-tubulin. Most of DsRed-CLIP-170 accumulates at the tip of microtubules. To understand the relation between microtubule dynamics and CLIP-170, we need to examine and calculate the dynamics of microtubules under these conditions. We have not yet performed such experiments.

Borisy: I think the point is still not clarified. What Richard Vallee is saying is that there is an anomaly between what you have reported for the persistence of CLIP on stable microtubules—that is, microtubules stabilized by interaction at the cortex—and what is known about the properties of microtubules in the cytoplasm. If a microtubule is stabilized in the cytoplasm with Taxol treatment, CLIP will dissociate very quickly. You have presented results suggesting that when the microtubule reaches the cortex in cells which have APC, CLIP remains. CLIP is remaining on a stabilized microtubule, whereas if a microtubule is stabilized in the cytoplasm it would not remain. This was the disparity that Richard Vallee was trying to clarify.

Vallee: My point was that there is an additional feature of this interaction, which is that one would assume that the affinity for the microtubule is constant and would remain the same. Even though the microtubule has run into IQGAP, CLIP-170 should dissociate from the microtubule. It doesn't seem to. This indicates that something else might be occurring in your experiments, such that you have changed the affinity of CLIP-170 for the microtubule end. This is extremely interesting, because you have converted CLIP-170 into a stabilizing protein, whereas it is normally not.

Nelson: You make the distinction between cytoplasmic microtubules and others. Isn't there also a distinction between microtubules which are intimately associated with the plasma membrane and those which aren't? They may have different properties.

Borisy: Yes, but Richard is trying to get to a molecular mechanism that explains the difference. He summarizes it by saying that what is implicit here is that APC is causing a change in the dissociation.

Chang: Do you need actin to stabilize the microtubules?

Kaibuchi: Yes. In our assay systems actin is needed for the accumulation of IQGAP1. If we treat the cells with cytochalasin D, IQGAP1 can't accumulate at the leading edge.

Firtel: Are there any other functions of IQGAP in these cells? In *Dictyostelium* the microtubules are much less important for migration. IQGAP appears to be involved in some aspect of the cortical tension. Have you seen any other effects of IQGAP on these cells, or do you think that the major effect is through the microtubule organization?

Kaibuchi: I think that IQGAP1 plays a critical role in the stabilization of actin filaments at the leading edges, and then links microtubules to actin filaments. IQGAP is also localized at the cell–cell adhesion site. So, I think IQGAP1 is involved in the stabilization of actin filaments at cell–cell contact sites.

Chang: Could the actin bundling activity be causing the multiple leading edges? If you overexpress bundling protein, does this cause multiple leading edges?

Kaibuchi: I am not sure, but I don't think so.

Gundersen: You are making the distinction between short-term and long-term stabilization of microtubules. Have you looked in APC or IQGAP knockouts for more long-term stabilized microtubules?

Kaibuchi: No, we have not yet looked at it. CLIP-170 is stabilized for up to two minutes under the physiological conditions at the leading edges. EGF can induce a leading edge with lamellipodia, where most of the CLIP-170 is stabilized at the leading edge. We have not examined the effects of Rho or extracellular signals which activate Rho.

Gundersen: I am wondering about the model that the transiently stabilized microtubules (by CLIP-170) may be precursors to long-lived microtubules, or maybe the transiently stabilized microtubules have a different function.

Kaibuchi: I think they may have a different function. We have no evidence indicating that microtubules which are stabilized by Rac/IQGAP1 are converted to more stabilized microtubules. As far as we observe the dynamics of CLIP-170 or tubulin, most of microtubules at the leading edges appear to be stabilized shortly and retracted (disassembled) again.

Gundersen: In our hands the long-term stabilized microtubules don't depend on Rac or Cdc42.

Kaibuchi: I agree. At the beginning I thought that Rac/Cdc42-induced stabilization of CLIP-170 might be an initial event for making more stabilized microtubules. But, now I think that these represent different pools of microtubules.

Nelson: We can isolate basal membrane patches from epithelial cells and find a beautiful lawn of microtubules, because these cells have been polarized and the plus ends of polarized microtubules are splayed out on the basal membrane. We have found that APC is localized along those microtubules. If we depolymerize microtubules, the APC spots stay behind on the membrane. If we add back fluorescently labelled tubulin dimer above the critical concentration, microtubules will re-polymerize to form a network over the APC spots. We argue that APC may be a component involved in stabilizing those microtubules and moreover there is APC on the plasma membrane 'constitutively', not just arriving there at the ends of microtubules. We think that APC is there at all times and is regulating microtubule organization on the basal membrane. When we look at microtubule dynamics within these patches we see a network where there are what look like cross-bridges between microtubules. These are very stable. The microtubules at these cross-bridges polymerize into space. As soon as these polymerizing microtubules hit an APC spot they are positioned and stabilized at that point. I would argue that APC plays additional roles to form a template on which microtubules are organized.

Kaibuchi: I agree. I think that APC also shows different localization from plus ends of microtubules. It is possible that APC localizes along microtubules and stabilizes them. APC also localizes at the cadherin-mediated cell–cell contact sites presumably through the interaction with β-catenin. I think that APC may stabilize microtubules at the cell–cell contact sites.

Regulation of microtubules by Rho GTPases in migrating cells

Gregg G. Gundersen, Ying Wen, Christina H. Eng, Jan Schmoranzer, Noemi Cabrera-Poch, Edward J. S. Morris, Michael Chen and Edgar R. Gomes

Departments of Anatomy & Cell Biology and Pathology, Columbia University, 630 W. 168th Street, New York, NY 10032, USA

Abstract. Microtubules (MTs) contribute to cell polarization and migration, but the molecular mechanism involved are unknown. We have explored signalling pathways that generate specific changes in MTs arrays in wounded monolayers of fibroblasts. In earlier work, we found that Rho GTPase and its effector mDia, stimulate selective MT stabilization in the lamella, whereas Cdc42 and the MT motor protein dynein regulate MT organizing centre (MTOC) reorientation towards the leading edge. We have now found that the MT tip proteins EB1 and adenomatous polyposis coli protein (APC) function with mDia to stabilize MTs and interact directly with mDia. EB1, APC and mDia localize to the ends of stabilized MTs suggesting that they may contribute to capping of these MTs. Models of MTOC reorientation suggest that the MTOC moves in front of the nucleus by dynein pulling on MTs. In contrast, we find by directly imaging MTOC reorientation that the nucleus moves rearward while the MTOC remains stationary. Rearward nuclear movement is coupled to retrograde actin-myosin flow and is regulated by Cdc42 and its effector myotonic dystrophy kinase-related Cdc42-binding kinase. Dynein is not involved in nuclear movement, but is essential to maintain the MTOC at the cell centroid. These results show that there are two Cdc42 pathways that regulate MTOC reorientation.

2005 Signalling networks in cell shape and motility. Wiley, Chichester (Novartis Foundation Symposium 269) p 108–126

Microtubules (MTs) are major elements of the cytoskeleton that work in conjunction with actin microfilaments to carry out most of the cell's motile activities including the generation and maintenance of asymmetric cell shape. A key feature of the behaviour of MTs is dynamic instability, which is their intrinsic ability to grow and shrink by alternatively adding and loosing subunits of tubulin. Dynamic instability has been proposed as a mechanism by which MTs probe the three-dimensional space of the cell to find sites of stabilization that have been activated by external signals (Kirschner & Mitchison 1986). By this model, MTs accumulation at sites of stabilization generates an asymmetric array of MTs. The stabilized MTs become post-translationally modified, which allows them to be distinguished from dynamic MTs, and this further enhances their capacity to generate cellular asymmetry.

Over the years, evidence has accumulated that MTs do indeed become stabilized and post-translationally modified at the onset of morphogenetic events. For example, during muscle cell differentiation, stabilized MTs selectively form along the axis of alignment of muscle cells before they fuse into multi-nucleated myotubes (Gundersen et al 1989). MTs are also stabilized early during neurite outgrowth and during epitheliogenesis (Bre et al 1990, Baas & Black 1990, Bulinski & Gundersen 1991). During asymmetric cell division in budding yeast, MTs are stabilized at bud sites and this contributes to the positioning of the spindle and nucleus during division (Adames & Cooper 2000). When fibroblasts migrate into an *in vitro* wound, stabilized MTs are selectively formed in the lamella and are oriented in the direction of cell migration (Gundersen & Bulinski 1988). These examples show that MT stabilization occurs during cell division, migration and differentiation.

Despite the wealth of evidence for selective stabilization of MTs during the establishment of cell asymmetries, our knowledge of the signals and structural molecules involved in selectively stabilizing MTs has lagged behind. Our laboratory and a number of others have used wounded monolayers of cells to search for these molecules. Migration of cells into such *in vitro* wounds requires MTs and the cells at the edge of the wound behave synchronously making this a useful system to study questions concerning MT rearrangements and the signalling pathways regulating them.

Two rearrangements of MTs during migration into the wound

We have characterized two major rearrangements of MTs that occur in wounded monolayers of fibroblasts:

- selective stabilization of MTs in the leading lamella and
- reorientation of the MT organizing centre (MTOC) to a position between the leading edge and the nucleus (Fig. 1).

Selectively stabilized MTs are thought to be formed by capping of their plus ends, because unlike dynamic MTs, the ends of stabilized MTs neither add nor lose tubulin subunits for considerable periods of time (Infante et al 2000). MTOC reorientation was originally described in endothelial cells, but is now known to occur in a wide range of adherent, migrating cells including fibroblasts, astrocytes and neuronal cells. Both events polarize the MT array: selective MT stabilization does so by generating a subset of modified MTs that are selectively directed toward the leading edge, while MTOC reorientation does

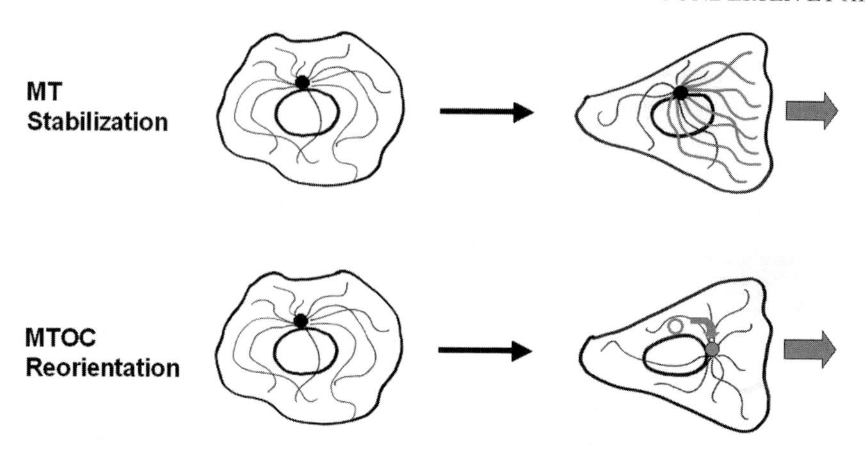

FIG. 1. Two rearrangements of MTs in migrating cells. Cells are shown before migration (left) and after stimulation of migration (right). Selective stabilization of MTs occurs in the lamella to generate a subset of post-translationally modified MTs oriented in the direction of migration (arrow). MTOC reorientation results in the positioning of the MTOC between the nucleus and the leading edge.

so by positioning the entire MT array in front of the nucleus. Selective MT stabilization and MTOC reorientation are initiated within minutes of wounding and reach maximum levels (on a cell-to-cell basis) by 1–2 h. Both of these processes may contribute to overall cell polarization by directing traffic of membrane precursors and other components to the leading edge (Bergmann et al 1983). Post-translationally modified MTs are better substrates for kinesin motor proteins (Liao & Gundersen 1998, Kreitzer et al 1999) and MTOC reorientation repositions the Golgi apparatus so that it is directed toward the leading edge of the cell (Kupfer et al 1982).

Identification of serum factors and signalling pathways involved in MT stabilization and MTOC reorientation

When serum-starved monolayers are wounded, neither selective MT stabilization or centrosome reorientation occurs (Gundersen et al 1994, Cook et al 1998, Palazzo et al 2001a), indicating that a serum factor(s) is necessary for both MT rearrangements. We identified this factor as the mitogenic lipid lysophosphatidic acid (LPA), which was both necessary and sufficient for serum-stimulated MT stabilization and is present in serum at levels consistent with the activity of serum (Nagasaki & Gundersen 1996, Cook et al 1998). LPA is also sufficient for stimulating centrosome reorientation, and serum LPA can account for the activity of serum. However, there may be additional factors in serum that can stimulate centrosome reorientation (Palazzo et al 2001a).

LPA acts on cells through a seven-span transmembrane G protein-coupled receptor to trigger a number of signalling pathways in cells. The pathway important for MT stabilization involves the activation of the small GTPase Rho (Cook et al 1998). Constitutively active alleles of Rho, V14Rho or L63Rho, are sufficient to induce stable MTs in the absence of serum or LPA, whereas the Rho-specific inhibitor, botulinum C3 toxin, blocks LPA-stimulated stable MT formation. Activation of this pathway stabilizes MTs within 5 minutes and recordings of the dynamics of the MTs shows that MTs near the cortex are captured and exhibit long pauses, while the dynamic instability of the bulk MTs is unaffected (Cook et al 1998).

Rho GTPases regulate cellular processes by interacting in their active GTP-bound state with effector molecules. For Rho there at least a dozen different effectors, including kinases and adaptor molecules. Using a Rho effector domain screen, we identified the formin, mDia2, as the likely candidate for mediating Rho effects on MTs (Palazzo et al 2001b). mDia like other formins contains two formin homology domains that interact with other molecules, an N-terminal GTPase binding domain and a C-terminal mDia autoinhibitory domain ('DAD'). We used constitutively active construct of mDia2 as well as the isolated DAD, which activates the endogenous mDia1 in our cells (Alberts 2001), to show that mDia activation was sufficient for inducing stable MT formation (Palazzo et al 2001b). mDia collaborates with another Rho effector, Rho kinase, to induce stress fibres. However, we found that Rho kinase was not necessary for MT stabilization. Below, we describe how mDia interacts with MT tip proteins to bring about MT stabilization.

LPA activates a second GTPase, Cdc42, to trigger MTOC reorientation in starved fibroblasts (Palazzo et al 2001a). Introduction of constitutively active Cdc42 alleles is sufficient to trigger centrosome reorientation in the absence of LPA. Cdc42 has also been implicated in MTOC reorientation of astrocytes in wounded monolayers (Etienne-Manneville & Hall 2001) and in endothelial cells subjected to shear stress (Tzima et al 2003). In both fibroblasts and astrocytes, the MT motor protein dynein and its regulator dynactin, appear to function downstream of Cdc42 to reorient the MTOC (Palazzo et al 2001a, Etienne-Manneville & Hall 2001).

Comprehensive screens to identify Cdc42 effectors involved in MTOC reorientation have not been conducted. However, the Cdc42 effector Par6, which was originally identified in a screen for proteins involved in *Caenorhabditis elegans* asymmetric cell division and its binding partner, PKCζ, have been implicated in MTOC reorientation of astrocytes, endothelial cells and fibroblasts (Etienne-Manneville & Hall 2001, Tzima et al 2003). In the work described below, we have identified a new Cdc42 effector, myotonic dystrophy kinase-related Cdc42-binding kinase (MRCK), that functions in a pathway separate from that of Par6-PKCζ and dynein to bring about MTOC reorientation.

Serum or LPA trigger both MT stabilization and MTOC reorientation and an important question is whether these two alterations are linked. Using inhibitors and activators of each process, we found that MT stabilization and MTOC reorientation were regulated independently and that neither process was required for the other (Palazzo et al 2001a). Thus, at least for these two MT rearrangements, there is no cross-talk between Rho and Cdc42 GTPases, and MT stabilization is not required for MTOC reorientation and vice versa.

Microtubule tip proteins function in the Rho-mDia pathway for selective MT stabilization

A comparison of the MT stabilization pathway in fibroblasts and a related stabilization pathway in yeast shows that both pathways rely on Rho GTPases and formins (mDia in fibroblasts; Bni1 in yeast) to regulate cortical stabilization of MTs (Fig. 2). In yeast, these MTs contribute to nuclear orientation and cell division by positioning the nucleus in the bud neck (Carminati & Stearns 1997, Bloom 2000). Yeast MTs are only transiently stabilized at cortical sites, yet the similarities in the two pathways suggest that additional aspects of the pathways may be conserved. In the yeast pathway, genetics, biochemistry and localization studies have established additional proteins that function in this pathway (Fig. 2). Two of these proteins, Kar9 and Bim1 are members of the class of MT tip proteins that reside primarily on the ends of growing MTs (Carvalho et al 2003). We tested whether homologues of the yeast tip proteins might be involved in the selective stabilization pathway in fibroblasts.

EB1 is the mammalian orthologue of Bim1 and the mammalian tumour suppressor adenomatous polyposis coli protein (APC) may be the functional homologue of Kar9, because like Kar9 it interacts with EB1 and there is limited sequence homology between the two proteins (Bienz 2001). We first tested whether EB1 was necessary for LPA-induced MT stabilization by knocking down EB1 levels by siRNA. Starved 3T3 cells treated with EB1 siRNA did not generate stable MTs in response to LPA stimulation (Wen et al 2004). EB1 siRNA reduced EB1 proteins levels as detected by immunofluorescence in 3T3 cells and Cos-7 cells and by Western blot of Cos-7 cell extracts (Wen et al 2004). These results indicate that EB1 is necessary for stabilization of MTs.

To further test the role of EB1 in MT stabilization, we took a dominant negative approach. EB1 has an N-terminal MT binding domain and a C-terminal domain that interacts with APC and the p150 Glued subunit of dynactin. We found that a GFP-tagged C-terminal fragment (EB1-C-GFP) prevented LPA or mDia-stimulated MT stabilization (Fig. 3) and that GFP-tagged full length EB1 stimulated MT stabilization when expressed in untreated starved 3T3 cells or

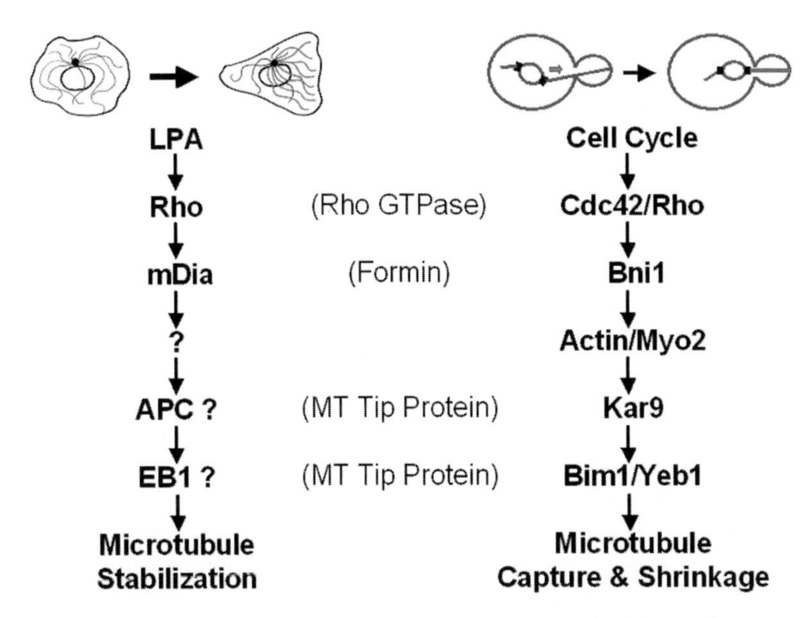

FIG. 2. Comparison of MT stabilization pathways in migrating fibroblasts and yeast.

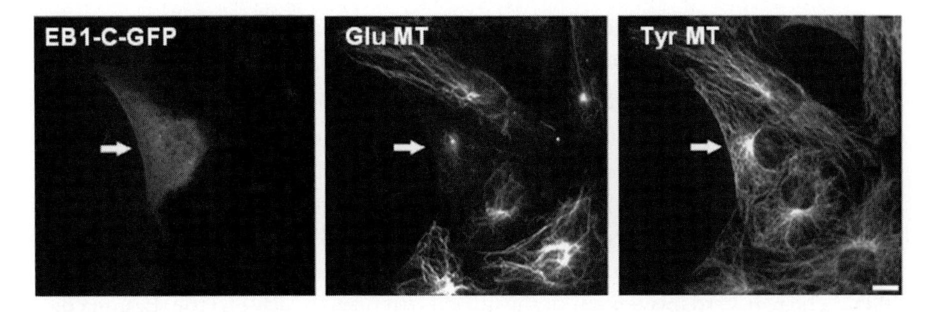

FIG. 3. EB1C-GFP acts as a dominant-negative to inhibit the formation of stable MTs induced by LPA. Expression of EB1C-GFP in a starved, wound-edge 3T3 fibroblast blocks formation of stable MTs induced by LPA. Arrow indicates cell expressing EB1-C-GFP. Stable MTs were assessed by immunofluorescently staining for post-translationally detyrosinated tubulin (Glu tubulin). Dynamic MTs were stained by an antibody to Tyr tubulin. Note that the cell expressing EB1-C-GFP does not contain stable Glu MTs, but has a normal array of dynamic MTs. Reproduced from Wen et al (2004) with permission. Bar, 10 μm.

starved 3T3 cells treated with C3 toxin to inhibit Rho (Wen et al 2004). Thus, EB1 functions downstream of Rho and mDia in the MT stabilization pathway.

Interactions studies showed that EB1-C-GFP interacted with APC *in vitro* and prevented the binding of endogenous EB1 and APC (Wen et al 2004). Also, APC overexpression, but not p150Glued overexpression, induced stable MTs in serum-starved 3T3 cells. These results suggest that EB1's interaction with APC is important for stable MT formation.

We next tested whether EB1 and APC interact with mDia. GST-tagged N-terminal EB1 fragment and APC-C pulled down an expressed active fragment of mDia. No actin or tubulin was detected in the pulldowns suggesting that both proteins can interact with mDia (Wen et al 2004). Recombinant proteins showed that these interactions were direct. Endogenous APC coimmunoprecipitated both EB1 and mDia showing that the endogenous proteins interact. As the domains of EB1 and APC that interact with each other are distinct from the domains that interact with mDia, it is possible that all three proteins can form a complex.

Attempts to localize APC or EB1 at the ends of stabilized MTs with conventional epifluorescence microscopy have been unsuccessful (Infante et al 2000). Using total internal reflection fluorescence (TIRF) microscopy, which has a much higher signal-to-noise ratio, we detected EB1 and to a lesser extent mDia1 and APC on the ends of stabilized MTs in immunofluorescently stained preparations (Wen et al 2004). Shifting controls showed that these colocalizations were not due to chance. These results suggest that EB1, along with mDia and APC may directly participate in capping the stabilized MTs.

MTOC reorientation requires two Cdc42 signalling pathways — one to regulate nuclear movement and one to maintain the centrosome at the cell centre

The initial reorientation of the MTOC in migrating cells has not been imaged directly. We conducted live imaging studies of 3T3 cells stably expressing GFP-tubulin (3T3GFP-Tu cells). Surprisingly, when starved 3T3GFP-Tu cells were stimulated with LPA (or serum) to reorient their MTOC, the nucleus moved rearward while the centrosome remained at the cell centroid (Fig. 4) (Gomes et al 2005). The rearward movement of the nucleus occurred relative to the leading edge at a rate of 0.28 μm/min, similar to the rate of retrograde actin flow. Using a fixed cell assay to measure MTOC and nuclear position, we found that two inhibitors of actin retrograde flow, cytochalasin D and blebbistatin, blocked rearward nuclear movement and MTOC reorientation, without affecting MTOC centration (Gomes et al 2005). Actin speckle analysis showed that LPA stimulated rearward actin flow.

Previous studies have implicated Cdc42 in MTOC reorientation and we examined whether Cdc42 also regulated nuclear position. Constitutively active L61Cdc42 stimulated rearward nuclear movement, while dominant negative N17Cdc42 or the CRIB domain of Pak1 both blocked LPA-stimulated rearward nuclear movement (Gomes et al 2005). There are two Cdc42 effectors, Pak1 and MRCK, that could be involved in regulating rearward nuclear movement through their abilities to regulate myosin phosphorylation. A dominant-negative

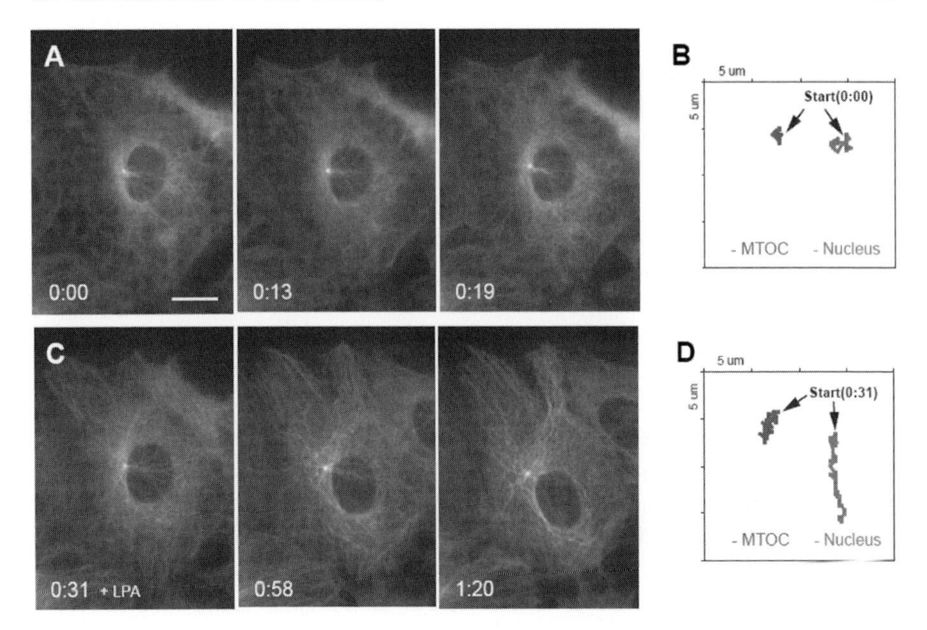

FIG. 4. LPA stimulates rearward nuclear movement to reorient the MTOC. 3T3GFP-Tu cells were starved and wounded and then movies were prepared before (A,B) and after LPA addition (C,D). Individual frames from the movie are shown in A,C (time is in h:min) and tracings of the position of the nucleus and MTOC are shown in B,D. Reproduced from Gomes et al (2005). Bar, 10 μm.

construct of Pak1 did not block rearward nuclear movement or MTOC reorientation. In contrast, two dominant-negative constructs of MRCK blocked rearward nuclear movement and MTOC reorientation stimulated by LPA. Expression of full-length MRCK alone stimulated rearward nuclear movement and MTOC reorientation. LPA, L61Cdc42 or wild-type MRCK stimulation of rearward nuclear movement and MTOC reorientation was accompanied by myosin activation as judged by increased phosphorylation of Ser19 of myosin light chain (Gomes et al 2005). These results show that there is a signalling pathway involving Cdc42, MRCK, myosin and actin that regulates rearward nuclear movement to reorient the MTOC.

Dynein and dynactin are involved in MTOC reorientation (Palazzo et al 2001b, Etienne-Manneville & Hall 2001) and are also known to regulate nuclear or MTOC position in many other systems (Reinsch & Gonczy 1998, Morris 2003). We tested whether dynein also played a role in rearward nuclear position. Inhibition of dynein with a function-blocking dynein intermediate chain (DIC) mAb (monoclonal antibody) did not prevent rearward nuclear movement as assayed by directly imaging living cells or with our fixed cell assay (Gomes et al

2005). However, the MTOC was not maintained at the cell centre and moved rearward with the nucleus in cells injected with DIC mAb and this prevented MTOC reorientation. Inhibitors of the Cdc42 effector Par6 and its binding partner, PKCζ, which had previously been implicated in MTOC reorientation (Etienne-Manneville & Hall 2001), also had no affect on rearward nuclear movement, but caused the MTOC to move rearward with the nucleus. Given that Par6, PKCζ and dynein all block MTOC reorientation with the same 'dynein' phenotype, this suggests that these proteins function in a pathway to maintain the MTOC in the cell centroid.

Conclusions

Our studies show how separate Rho and Cdc42 signalling pathways regulate two rearrangements of MTs in migrating cells. Despite the distinct molecules and separate effects these pathways have on MTs, the pathways share a number of features. Foremost, both are regulated by members of the Rho GTPase family and involve distinct GTPase effectors. Our results and those from other laboratories (Gundersen 2002, Gundersen et al 2004), indicate that MTs are major targets of Rho GTPases. Coupled with earlier work establishing that actin filaments are major targets of GTPases it is now becoming more clear that the cytoskeleton as a whole is the principal site of action of the Rho GTPases.

A second important point is that the pathways regulating MTs in migrating fibroblasts strongly resemble those in budding yeast. This similarity is particularly striking in the case of the MT stabilization pathways, where a Rho GTPase, a formin and MT tip proteins participate in both pathways (Fig. 1). Although it is too early to know whether these pathways are used to regulate MTs in other contexts, the diverse systems used and the involvement of the pathways in regulating MTs for cell division and cell migration, point to a more general function for these evolutionary conserved signalling pathways.

Finally, it is important to note that both MT responses occur in the context of cross-talk with the actin cytoskeleton. There is of course the fact that Rho GTPases trigger changes in both actin and MTs and hence there is coordinated stimulation of changes in both cytoskeletal elements. In the case of MT stabilization, the divergence in signalling to the MT and actin cytoskeletons seems to occur at the level of the formin, mDia. mDia is a well-known nucleator of actin filaments (Wallar 2003), but also has the capacity to interact with MT tip proteins and stabilize MTs. Neither of these activities appears to be necessary for the other, so one possibility we are exploring is whether there is competition between the cytoskeletal systems for the formin. In the case of MTOC reorientation, the target divergence also seems to occur at the level of the effector. MRCK is a well-established myosin kinase that results in myosin activation (Leung et al 1998). Yet, because it is sufficient to induce MTOC reorientation by both inducing rearward nuclear movement and maintaining the MTOC at the cell centroid, it is likely that it also regulates the dynein side of the branched pathway. MTOC reorientation is emerging as a clear case where the actin and MT cytoskeletons must be coordinated to establish polarity in cells.

An important question is how the actin–myosin rearward flow actually moves the nucleus. Is it just that rearward moving actin filaments push the nucleus rearward, much like a bulldozer? Or perhaps there are specific molecules that mediate a connection between the nuclear envelop and rearward moving actin cables, much like a cable car. Candidates for such linking proteins are the Syne/Anc family of proteins with binding domains for nuclear envelope proteins and the actin cytoskeleton (Starr & Han 2003). Another interesting question is whether this pathway functions in other cases of nuclear movement, such as differentiating muscle cells or in migrating neurons.

Acknowledgements

We thank B. Wallar and A. Alberts for help with aspects of the mDia story not directly mentioned here. This work was support by NIH grant GM062938 (to GGG).

References

Adames NR, Cooper JA 2000 Microtubule interactions with the cell cortex causing nuclear movements in *Saccharomyces cerevisiae*. J Cell Biol 149:863–874

Alberts AS 2001 Identification of a carboxyl-terminal diaphanous-related formin homology protein autoregulatory domain. J Biol Chem 276:2824–2830

Baas PW, Black MM 1990 Individual microtubules in the axon consist of domains that differ in both composition and stability. J Cell Biol 111:495–509

Bergmann JE, Kupfer A, Singer SJ 1983 Membrane insertion at the leading edge of motile fibroblasts. Proc Natl Acad Sci USA 80:1367–1371

Bienz M 2001 Spindles cotton on to junctions, APC and EB1. Nat Cell Biol 3:E67–68

Bloom K 2000 It's a kar9ochore to capture microtubules. Nat Cell Biol 2:E96–98

Bre MH, Pepperkok R, Hill AM et al 1990 Regulation of microtubule dynamics and nucleation during polarization in MDCK II cells. J Cell Biol 111:3013–3021

Bulinski JC, Gundersen GG 1991 Stabilization of post-translational modification of microtubules during cellular morphogenesis. Bioessays 13:285–293

Carminati JL, Stearns T 1997 Microtubules orient the mitotic spindle in yeast through dynein-dependent interactions with the cell cortex. J Cell Biol 138:629–641

Carvalho P, Tirnauer JS, Pellman D 2003 Surfing on microtubule ends. Trends Cell Biol 13:229–237

Cook TA, Nagasaki T, Gundersen GG 1998 Rho guanosine triphosphatase mediates the selective stabilization of microtubules induced by lysophosphatidic acid. J Cell Biol 141:175–185

Etienne-Manneville S, Hall A 2001 Integrin-mediated activation of Cdc42 controls cell polarity in migrating astrocytes through PKCzeta. Cell 106:489–498

Gomes ER, Jani S, Gundersen GG 2005 Nuclear movement by Cdc42-MRCK regulated actin-myosin flow establishes MTOC polarization in migrating cells. Cell 121:451–463

Gundersen GG 2002 Microtubule capture: IQGAP and CLIP-170 expand the repertoire. Curr Biol 12:R645–647

Gundersen GG, Bulinski JC 1988 Selective stabilization of microtubules oriented toward the direction of cell migration. Proc Natl Acad Sci USA 85:5946–5950

Gundersen GG, Khawaja S, Bulinski JC 1989 Generation of a stable, posttranslationally modified microtubule array is an early event in myogenic differentiation. J Cell Biol 109:2275–2288

Gundersen GG, Kim I, Chapin CJ 1994 Induction of stable microtubules in 3T3 fibroblasts by TGF-beta and serum. J Cell Sci 107:645–659

Gundersen GG, Gomes ER, Wen Y 2004 Cortical control of microtubule stability and polarization. Curr Opin Cell Biol 16:106–112

Infante AS, Stein MS, Zhai Y, Borisy GG, Gundersen GG 2000 Detyrosinated (Glu) microtubules are stabilized by an ATP-sensitive plus-end cap. J Cell Sci 113:3907–3919

Kirschner M, Mitchison T 1986 Beyond self-assembly: from microtubules to morphogenesis. Cell 45:329–342

Kreitzer G, Liao G, Gundersen GG 1999 Detyrosination of tubulin regulates the interaction of intermediate filaments with microtubules in vivo via a kinesin-dependent mechanism. Mol Biol Cell 10:1105–1118

Kupfer A, Louvard D, Singer SJ 1982 Polarization of the Golgi apparatus and the microtubule-organizing center in cultured fibroblasts at the edge of an experimental wound. Proc Natl Acad Sci USA 79:2603–2607

Leung T, Chen XQ, Tan I, Manser E, Lim L 1998 Myotonic dystrophy kinase-related Cdc42-binding kinase acts as a Cdc42 effector in promoting cytoskeletal reorganization. Mol Cell Biol 18:130–140

Liao G, Gundersen GG 1998 Kinesin is a candidate for cross-bridging microtubules and intermediate filaments. Selective binding of kinesin to detyrosinated tubulin and vimentin. J Biol Chem 273:9797–9803

Morris NR 2003 Nuclear positioning: the means is at the ends. Curr Opin Cell Biol 15:54–59

Nagasaki T, Gundersen GG 1996 Depletion of lysophosphatidic acid triggers a loss of oriented detyrosinated microtubules in motile fibroblasts. J Cell Sci 109:2461–2469

Palazzo AF, Joseph HL, Chen YJ et al 2001a Cdc42, dynein, and dynactin regulate MTOC reorientation independent of Rho-regulated microtubule stabilization. Curr Biol 11:1536–1541

Palazzo AF, Cook TA, Alberts AS, Gundersen GG 2001b mDia mediates Rho-regulated formation and orientation of stable microtubules. Nat Cell Biol 3:723–729

Reinsch S, Gonczy P 1998 Mechanisms of nuclear positioning. J Cell Sci 111:2283–2295

Starr DA, Han M 2003 ANChors away: an actin based mechanism of nuclear positioning. J Cell Sci 116:211–216

Tzima E, Kiosses WB, del Pozo MA, Schwartz MA 2003 Localized cdc42 activation, detected using a novel assay, mediates microtubule organizing center positioning in endothelial cells in response to fluid shear stress. J Biol Chem 278:31020–31023

Wallar BJ, Alberts AS 2003 The formins: active scaffolds that remodel the cytoskeleton. Trends Cell Biol 13:435–446

Wen Y, Eng CH, Schmoranzer J et al 2004 EB1 and APC bind to mDia to stabilize microtubules downstream of Rho and promote cell migration. Nat Cell Biol 6:820–830

DISCUSSION

Kaibuchi: What percentage of MTs can be stabilized for a long time?

Gundersen: On the basis of the amount of post-translationally detyrosinated tubulin (Glu tubulin), it is around 10–15% in fibroblasts. This is also consistent with the percentage of MTs that exhibit long term pausing in cells with activated Rho. In differentiated tissues, such as the brain, it may be more like 50%.

Kaibuchi: During cell migration, my guess would be that most of the MTs appear to be stabilized at the leading edge for up to 2 min, but not in other areas such as opposite sites of the leading edge. So, you think that 10–15% of MTs are capped by Glu tubulin in migrating cells.

Gundersen: We distinguish between Glu tubulin and Glu MTs. Even a dynamic MT will have a low level of Glu tubulin in it, but because it turns over so rapidly it never accumulates Glu tubulin to a high degree. Glu MTs are those that have been stabilized for long enough that they accumulate substantial levels of Glu tubulin and are detected by immunofluorescence with antibodies to Glu tubulin. It is a restricted subset of MTs that accumulate a lot of this.

Kaibuchi: Do you think that most of the stabilized MTs are directed to focal contacts or adhesions?

Gundersen: We have shown many times that very few of the stabilized MTs have their ends localized in the focal adhesions (for example, see Infante et al 2001). It would have been a very nice situation if the stable MTs went right to the focal adhesion. They are close, but they are not actually in the focal adhesion.

Humphries: How do you explain that they are integrin dependent?

Gundersen: Focal adhesions are signalling centres. When a new focal adhesion is formed, perhaps the activated signalling pathways activate factors necessary for MT stabilization. One factor that appears to be involved is the GM1-containing lipid raft that forms at the leading edge and is necessary for formation of stable MTs. It looks like integrins have a significant effect on trafficking select things. We don't know whether or not their affect on GM1 lipid rafts is an effect on endocytosis or an effect on exocytosis.

Humphries: Are you sure it is an integrin? You are using whole fibronectin.

Gundersen: We have also done this on vitronectin and collagen, so it appears stable MTs require integrin engagement, and that a number of integrins have similar effects on stable MT formation. Beyond this, no. But it is clearly also FAK dependent, which gives us more confidence that it is due to integrin signalling.

Sheetz: In terms of the myosin-dependent effect, is this a contraction of the back part of the cell? Do you see the rear portion of the cell contracting?

Gundersen: We don't see a big contraction. It is an interesting point: in the few other systems where nuclei are dispersed by the actin cytoskeleton — for example, the syncytial blastoderm of *Drosophila* embryos — there is an actual contraction that drives redistribution of nuclei. In this case it is not clear whether the nuclei are moving because of the contraction or are actually moving on their own. We don't see a distinct contraction, though. We see a smooth flow. Our working models range from a 'bulldozer model', where the rearward actin flow pushes the nucleus back, to a maybe more interesting 'cable car model' where there are actually proteins that hook the nucleus up to the rearward-moving actin cytoskeleton.

Drubin: How stable are these MTs? Can you speculate on their function?

Gundersen: The best way to answer this is their longevity. In the TC-7 monkey kidney cells where we measured it, you can see MTs that are stable for 16 h or virtually the entire cell cycle. It is a dwindling population, though, so the half-life may be about 4 h in those cells. In the migrating 3T3 fibroblasts we haven't gone back and looked at the half-life. I suspect that they are not quite as old. Another relevant point is that it takes about 30 min for them to accumulate enough Glu tubulin for us to see them by immunofluorescence. This is the lower limit of how old they are. We think that they are formed very quickly because if we use other measures we can see stabilization in serum-starved 3T3 fibroblasts within 5 min of LPA treatment. Functionally, we have shown that the Glu tubulin seems to interact better with kinesin. One thought is that they are better tracks, or specialized tracks, for the transport of material to the front of the cell.

Peter: Have you looked in *Caenorhabditis elegans* where there are dramatic nuclear movements in the early embryo that are Cdc42-dependent.

Gundersen: Which ones aren't Par dependent?

Peter: The migration of the nuclei into the middle of the cells occurs after fertilization, but before the asymmetric cleavage is set up.

Gundersen: That's an interesting idea, but we haven't looked. We have only very recently found evidence that MRCK is involved in the nuclear movement. I don't know whether *C. elegans* even has an MRCK.

Cai: It seems to me that it is the relative distance between the leading edge and the MTOC that is the important parameter a cell needs to adjust in the process of directional transport. Why is the rearward movement of the nucleus required for establishing the polarity?

Gundersen: At a minimum, it places the MTOC in front of the nucleus relative to the leading edge. Since the Golgi is localized at the MTOC, the preferential localization of the MTOC relative to the leading edge may contribute to delivery of membrane precursors and other factors towards the leading edge. Moving the nucleus rearward may also be important and we are trying to test this.

Cai: But the position of the MTOC hasn't changed.

Gundersen: That's correct. The MTOC seems to be maintained in the centroid by a rather extensive involvement with dynein, dynactin, Pars and things like this. That is clearly an important function. Why the nucleus is moved rearward is anyone's guess at this point. If you look at any MT array in a tissue culture cell and look behind the nucleus from the MTOC the density of MTs there is much less: the MTs have to go around the nucleus. This may be one way of giving the MTs, through their dynamics, more easy access to the front of the cell.

Vallee: As your collaborator I tend to take a more dynein-centred view of the world. Pulling on the MTs might be the important issue, and the nucleus is moving back passively. I think this is an equally reasonable hypothesis. The constant cortical dynein is keeping the centrosome at the centre of the cell.

Gundersen: I wouldn't use the word 'passive', since it seems to involve actin, MRCK and myosin. I think what you are getting at is whether it is significant, and we just don't know this at the moment. Is the rearward position of the nucleus in some way important for cell migration? There are many cases in biology where position of the nuclei in epithelial cells is critical for the function of the epithelial cell. In *Drosophila* optic epithelium there are mutations that will cause the nuclei to stay at the bottom of the cell. If they don't go up to the centre this interferes with all kinds of polarity at the apical surface.

Chang: Do you think your MTOC is being pulled by dynein from the cortex? Or is the dynein working at the MTOC?

Gundersen: It is much easier to explain how it stays in the centre if dynein is pulling from the cortex. But we can't rule anything out yet. One observation we have made that may be relevant is that when LPA is added, the MTs appear to become more rigid. We are trying to understand whether this is evidence for them interacting with cortical sites very rapidly.

Chang: Do you see any MTs that are wrapping around the cortex?

Gundersen: We don't have any evidence for dynein pulling on a MT yet. This doesn't mean it doesn't happen. There are many MTs that will bend when they encounter the cell edge and these then tend to grow parallel to the leading edge of the cell. These tend to get swept rearward by that centripetal flow. Whether these parallel MTs are somehow hooking up to the actin cytoskeleton, or they are just trapped in a rearward flow of actin, we do not yet know.

B Lane: There are a large number of tissues in which there is this nuclear movement when the cells go through the cell cycle. As the cells go into mitosis the nucleus shifts substantially through the cytoplasm. It might be worth thinking about those systems as well.

Gundersen: Yes. There is a great example in the neuroepithelium where there is movement of nuclei during the cell cycle—when the nuclei return to the basal aspect of the cell, they undergo mitosis.

B Lane: It is also seen in several embryonic epithelia.

Titus: Where is the myosin II and MRCK localized in these cells?

Gundersen: We haven't looked carefully at this yet, although we have just got an antibody to MRCK. Work from Thomas Leone has shown that when MRCK is activated it tends to be recruited to the cortex.

Nelson: Another example of MTOC movement is activation of T cells by B cells. There the MTOC moves round to the point of contact.

Gundersen: This is why we thought the MTOC itself was going to move. In yeast, *Aspergillus* and T cells the MTOC undergoes a reorientation and in those cases it does actually move.

Nelson: In the context of evolutionary conservation of pathways, is there any way that the pathways that you have enumerated might be involved in that MTOC movement independent of any nuclear movement?

Gundersen: Yes. In our cells, if you ask whether dynein is going to be pulling on something you could imagine various regulatory factors which would turn on or turn off dynein at various sites. Differential regulation of dynein may lead to a position of the MTOC away from the centroid.

Borisy: The most dramatic example of nuclear movement is pronuclear fusion. This is motor driven.

Manser: I have a couple of comments in relation to MRCK. First, it is regulated by diacylglycerol (DAG). There will be a very interesting second messenger pathway connection. Second, we noticed for a number of years that when we put hyperactive versions of MRCK into cells we see a massive distortion of the nucleus. This would fit with there being some close connection between the actin and the nucleus.

Vallee: I am realizing that I have a bias about the centrosome issue. Universally, I think the centrosome is under the control of cortical dynein through the MTs that go out into the cell cortex. There is evidence that nuclear movements in *Aspergillus* are mediated by cortical dynein. I could cite more examples. Perhaps the difference between the T cell situation and the fibroblasts is that in fibroblasts the centrosome is already at the very centre of the cell. This is the reason why they are different from the T cells. The centrosome in the fibroblast has nowhere it needs to move: for some reason the cell keeps the centrosome just where it is as it activates the motile machinery. Everything starts being swept to the back of the cell but dynein holds the centrosome where it should be. This is the same mechanism as in the T cell, but it looks different to the investigator.

Borisy: Why does it look different?

Vallee: In the T cell the MTOC is random, so when dynein becomes activated at the cortex this will lead to a movement of the centrosome. In the fibroblasts the centrosome sits at the very centre of the cell, and dynein is responsible for that. In the wound-healing system one could argue that dynein becomes activated and the MTs are pulled, but the centrosome is at its destination already.

Borisy: That makes sense. In Greg Gundersen's experiments when he injects the antibody against dynein, the centrosome gets swept back. The continued action of dynein would be necessary to maintain the position of the centrosome in the presence of a rearward or contractile force that repositions the nucleus.

Sheetz: My understanding of this is that the dynein that may be relevant is on the lower surface, not necessarily that at the leading edge.

Vallee: The dynein would be everywhere. A very interesting question in the wound-healing situation is that now you have imbalance of forces whereas before you had a balance of forces, so how do dynein, myosin and actin deal with that situation?

Gundersen: That's a good question that we now must address. I want to point out something else. In many cells the MTOC is attached in some way to the nucleus. In our experiments, when the nucleus is moving rearwards, the MTOC comes off it a little bit. So there may be a further step in here that is also regulated, and it could be that the key event is to get the MTOC off the nucleus, so only the nucleus is subjected to this rearward flow.

Chang: Our thinking may be too simplistic. There are plenty of data showing that many MTs are also attached to this actin flow. Dynein is involved in attaching the centrosome to the nuclear envelope in many cases. There could be multiple things happening that are too complicated for us to appreciate at the moment.

Gundersen: In has been our experience studying MTOC reorganization, which seems like a rather simple reorganization, that every time we look at it we are surprised by something new that comes up. So I agree with you: it may be very involved to just position the nucleus with respect to the MTOC.

Firtel: We have MTs going to the leading edge. The question is: what are they doing? One possibility is that they are transporting vesicles. One thing we haven't discussed is that the leading edge needs to push forwards, so we need to deal with the membrane issue. Mark Bretcher did a nice experiment in *Dictyostelium* in which he made a temperature-sensitive mutant of NSF. Under these conditions actin can be polymerized but the membrane can't be protruded. The idea here is that you need to have accumulation and a lot of movement of vesicles to the membrane in order to expand the surface. Is this a possible use of the MTs? What is being transported on them and why are they needed at the leading edge?

Sheetz: We have a lot of evidence that there is an excess of membrane on cell surfaces. There is very little resistance to movement of membrane from one place to another. You can pull on the membrane with laser tweezers and pull out significant portions with relatively little input of energy.

Gundersen: You may not need extra membrane per se, but that does not exclude the possibility that there may be a need to deliver membrane to balance or counteract endocytosis.

Sheetz: In terms of active components, enzymes and so on, you could argue that proteins which are degraded at the membrane surface may be depleted in certain regions and therefore need to be transported.

Kaibuchi: I think that some adhesion molecules and cytoskeletal components such as APC may be continually transported along MTs to the leading edge for making new lamellipodium and for its turnover.

Borisy: We have shifted gears from nuclear movement and MTOC reorientation to MT capture. We are asking questions about why we have to stabilize and capture MTs, and how this is related to polarization of the cell. Greg suggested that one purpose of selective stabilization of dynamic MTs is to then allow for transport of material to the leading edge, which results in polarization. This echoes previous work from John Singer's lab which showed selective transport of material to the leading edge. He used a viral assembly system. So there is good reason to be thinking about directed transport of material to the presumptive leading edge, even though Mike Sheetz's experiments say that it is not necessary from a biophysical point of view. Let's come back to the topic of MT capture. What evidence do we have for MT capture in mammalian cells? Are we defining capture as a MT staying at the edge of the cell for more than a few tenths of a second? Kozo Kaibuchi was saying that he didn't see MTs staying at the leading edge more than 2 min, and when Greg Gundersen was talking about stabilizing stable MTs they weren't MTs that necessarily reached the edge. They were stabilized by being capped at their plus end somewhere in the cytoplasm by a mechanism as yet unspecified.

Gundersen: And we don't know that these stable MTs stay in place. We are trying to find ways to look at the dynamics of stable MTs.

Borisy: What is the definition of MT capture?

Gundersen: Any interaction with a localized factor that changes the intrinsic dynamics of the MT. There may be different types of MT capture leading to different effects on MTs. For example, MT capture by the Rho-mDia-EB1-APC pathway I described leads to plus end stabilized MTs with a long half-life. The MTs captured by IQGAP-CLIP-170 are only transiently stabilized. And MTs that are captured by dynein may be pulled on and may not need to be stabilized.

Borisy: Up to this point, how much can we say about the change in dynamics?

Gundersen: For the long-lived stable MTs we study, we haven't been able to follow them yet from the moment of capture to the actual time the MT is capped. But the capture in this case is hours long.

Borisy: You know that MTs become stabilized, but you don't know how they became stabilized. We are asking whether or not they become stabilized by cortical capture specifically: you don't know that the Glu MTs have become stabilized by encountering the cortex.

Gundersen: We believe they are captured by encountering the cortex. In movies of fluorescently labelled MTs in cells in which Rho was activated, we detected a subset of MTs that grew to the cortex and then paused for long periods (Cook et al 1998). We only observed these long-paused MTs at the very edge of the cell. A second piece of evidence comes from our total internal reflection fluorescence (TIRF) localization of Glu MT ends. TIRF is limited to only the bottom ~100 nm of the cell and our study showed that many of the Glu MT ends are detectable by

TIRF (Wen et al 2004). This suggests that they are very close to the cortex at the bottom of the cell.

Alberts: Is this a problem of using different terms? You are using jargon that has been hijacked from the yeast field.

Gundersen: 'Capture' was actually hijacked from capture of MTs at the kinetochore. There is a lot of similarity between the players involved in cortical MT capture and what is going on at the kinetochore in terms of MT capture. In our case we don't have a full understanding of what happens to a MT when Rho is activated. When we look at the dynamics of the MTs we see very long paused MTs. They pause for up to 10 min, which is as long as we can follow them.

Borisy: That is a pause somewhere in the cytoplasm?

Gundersen: It is a pause right at the leading edge, presumably in the cortex. We haven't been able to follow them from that stage.

Kaibuchi: We cannot see any stabilization of CLIP-170 in the cytoplasm of the centre area of the cells. Also, we cannot see any stabilization of CLIP-170 near the plasma membranes without a leading edge. When the cell is stimulated with growth factors such as EGF, and a single leading edge with lamellipodium is induced. Most of CLIP-170 appears to be stabilized just beneath the cortical region at the leading edge. In other word, CLIP-170 appears to be captured just beneath the cortical region where lamellipodium is induced.

Chang: I have been trying to reconcile the results presented by Kozo and Greg. They seem to involve different mechanisms: Kozo's is based more on Rac and Cdc42, while Greg's is more Rho based. On the other hand, it seems that to get superstable MTs they first need to be stabilized for at least two minutes for an initial step. There could be an initial stabilization machinery and then something else to stabilize them longer.

Gundersen: I agree that is an attractive hypothesis. But if we interfere with Rac or Cdc42 it doesn't interfere with the stabilization we are looking at. The effect of Rac and Cdc42 may be independent of the effects of Rho on MTs. Indeed, in our experiments, active Rac and Cdc42 are not able to generate the long-lived stable MTs, whereas active Rho does. Also, inhibitors of Rac or Cdc42 have no effect on the stable MTs induced by the serum factor, LPA, whereas inhibitors of Rho block the LPA-induced stabilization of MTs. It may be that the mechanisms through which Rac and Cdc42 stabilize MTs are very different from that through which Rho stabilizes MTs. Also, it is likely that MTs stabilized by different mechanisms may perform different functions.

Alberts: Are they mutually exclusive?

Kaibuchi: At the level of activation of Rac and Rho, Rho is not subsequently activated by Rac. If you look at the activation of Rac or Cdc42 by using a fluorescence resonance energy transfer (FRET) system, we can see the activation of Rac/Cdc42 for less than 10 seconds. Their activation may not be followed by

Rho activation, because the activation area is different. I do not think that MTs, which are transiently stabilized by Rac or Cdc24, are subsequently stabilized by Rho, because most of MTs at the leading edge appear to be stabilized for a short period and then are retracted. This observation suggests that these MTs represent a different pool from those stabilized by Rho. But, I cannot neglect the attractive hypothesis that Fred has proposed. Also, I cannot say whether these two pools are mutually exclusive or not.

Gundersen: I would like to add that in our system, Rho clearly achieves its stabilizing affect on MTs independently of Rac and Cdc42.

Alberts: There are many examples of exchange factors that can exchange and activate Rho, Rac, or Cdc42 in combination. It could be that your probe isn't sufficient to detect the activation of Rho in that situation.

Sheetz: We haven't talked about the work from Vic Small's lab suggesting that MTs are involved in the disassembly of focal contacts. Is any of this relevant to that? The reason I was asking about contraction at the back is that I was wondering whether some of the actin at the rear of the cell might be depolymerizing, allowing the tail to move forward.

Kaibuchi: In this regard, we have not examined the dynamics of CLIP-170 at the cell to substratum contact sites.

Chang: What can the MTs do in 2 min? It is not enough time to direct vesicle transport.

Borisy: Let's restate Vic Small's observation. MTs are exploring and with some probability greater than random they encounter a focal adhesion. After the MT encounters the focal adhesion the focal adhesion will disassemble. The lifetime of the focal adhesion is affected by the encounter with the MT. MTs do a kiss-and-run: they touch the focal adhesion, there is some transfer of information, and as a consequence the focal adhesion disassembles or has increased probability to disassemble. This would be very significant for directed cell motion: if you disassemble adhesions in one place, there is opportunity for extension, and you remove forces that retain the cell in those positions.

Sheetz: The limiting factor for most fibroblast motility is lifting of the hind foot, rather than extending the leading edge.

Borisy: A lot of the discussion here has been on MTs exhibiting a positive effect for migration. In the context of being oriented towards the wound, and delivering material that will promote migration, this is a positive effect. But in the Small view, the MTs are having a negative effect on something that restrains motion. Small's view is that you promote disassembly of adhesions.

Sheetz: Delivery might be of enzymatic systems, which have an amplification factor. In other words, the movement of something to a region where it is going to be active and then causing a local focus of activity, would be a more favourable model than a simple mechanical effect.

Nelson: The other experiment worth mentioning here is from Waterman-Storer where she showed that expression of dominant-negative 'mutants' of PKD blocked migration of Swiss 3T3 cells at the leading edge (Prigozhina & Waterman-Storer 2004). She speculated that this was because it blocked a secretory pathway. At about the same time Malhotra's lab and ours showed that these PKD mutants block exocytosis of vesicles from the trans-Golgi network, likely those containing integrins thereby blocking their delivery to the leading edge (Seufferlein et al 2004).

Sheetz: Things don't deliver to the leading edge. The leading edge is a good 2 μm away from the closest MTs.

Nelson: I'll change that and say that delivery to a region that is actively involved in forward migration is important.

Firtel: You have contractility of myosin in the posterior which is coordinated with protrusion at the anterior and this is regulated by signalling pathways. On top of this highly regulated process, to have a random process where MTs are hitting focal adhesions doesn't make sense. Why not regulate the focal adhesion release by the same mechanism in the same way you are spatially organizing them with the contractility?

B Lane: Because you don't want all the junctions to be released at the same time. You can't have the junctions all catastrophically falling apart. Random hits will lead to a more controllable series of tissue changes.

Sheetz: These are stochastic processes rather than managed processes.

Firtel: I like managed processes. They are easier to understand.

Borisy: Who is the manager?

Firtel: Rho. Rho is modulating the kinetics of myosin contractility.

Gundersen: It is interesting that the release of focal adhesions is occurring in the tail where lots of people think Rho activity is the highest. This suggests that you cannot simply reverse Rho activation to turn over focal adhesions. Perhaps there is another way to turn them over in areas where Rho activity is high.

Borisy: Focal adhesions need to be released not only in the tail. If you are thinking about releasing focal adhesions in the tail you are thinking of a persistent directional motion. Now think in terms of turning of the cell. If you want to turn to the right, you need to release adhesions on the left side.

Sheetz: The standard fibroblast morphology is a broad lamellipodium at the front, and a much smaller cross-sectional area at the nucleus. Two thirds of the contacts must be released by the time the nucleus moves to that point. Thus, most of the release has to occur in the front of the cell.

Firtel: What you are suggesting is that the MTs' random interaction is modulating the total level of contact, and this keeps things flexible. This is a non-regulated stochastic type of regulatory pathway.

Borisy: It is probably regulated. In one sense it is regulated by the spatial distribution of MTs: where MT plus ends are more abundant, that is where you will have a greater probability of focal adhesion release. Factors that affect MT orientation would affect focal adhesion release and therefore directionality.

Nelson: In Vic Small's data, how much is correlation?

Borisy: We have confirmed all his results.

Nelson: He watches cells spread out in the presence of nocadazole and sees the focal adhesions are much more stable under those conditions. In another type of experiment he treats with nocadazole, washes it out and then watches MTs flood in to the edge of the cell. That edge retracts. Those are correlative events. Which experiment demonstrates this directly?

Gundersen: I don't think it has been done yet.

Braga: In his case, polarizing migration is not investigated. Perhaps that is a difference between his experiments and yours.

Gundersen: Yes.

References

Cook TA, Nagasaki T, Gundersen GG 1998 Rho GTPase mediates the selective stabilization of microtubules *in vivo*. J Cell Biol 141:175–185

Infante A, Stein M, Zhai Y, Borisy GG, Gundersen GG 2000 Detyrosinated (Glu) MTs are stabilized by an ATP-sensitive plus-end cap. J Cell Sci 113:3907–3919

Prigozhina NL, Waterman-Storer CM 2004 Protein kinase D-mediated anterograde membrane trafficking is required for fibroblast motility. Curr Biol 14:88–98

Seufferlein T, Mellman I, Nelson WJ, Malhotra V 2004 Protein kinase D (PKD) regulates basolateral protein exit from the trans-Golgi network. Nature Cell Biol 6:106–112

Wen Y, Eng CH, Schmoranzer J et al 2004 EB1 and APC bind to mDia to stabilize microtubules downstream of Rho and promote cell migration. Nat Cell Biol 6:820–830

Actin organization in the early *Drosophila* embryo

Eyal D. Schejter

Department of Molecular Genetics, Weizmann Institute of Science, Rehovot 76100, Israel

Abstract. Organization of the cortical cytoplasm during the syncytial blastoderm stages of early *Drosophila* embryogenesis relies on cyclic transitions between transient microfilament structures. Microtubule-organizing centres (MTOCs) appear to provide the instructive cues governing this dynamic, cell-cycle-dependent process. Using a genetic approach, we have identified key roles for two molecular pathways in mediating these events. The conserved Arp2/3 microfilament nucleation machinery, likely acting in response to the activating element SCAR, plays an essential role in establishment of a cortical F-actin array, and contributes to specific aspects of cyclic microfilament restructuring. Defective cortical microfilament organization is the primary phenotypic feature of embryos derived from mothers bearing mutations in the *sponge* locus. Several lines of investigation suggest that the primary defect in *sponge* lies in a faulty cortical microfilament response, downstream of the centrosomal signal. We have determined that *sponge* encodes a *Drosophila* homologue of the evolutionarily-conserved CDM (DOCK180) protein family.

2005 Signalling networks in cell shape and motility. Wiley, Chichester (Novartis Foundation Symposium 269) p 127–143

The earliest stages of *Drosophila* embryogenesis take place within a single, large cell and thus present a complex problem of cytoplasmic organization. Following fertilization, rapid and synchronous rounds of nuclear division proceed without intervening cytokinesis. Consequentially, the series of large-scale morphogenetic events characteristic of early embryogenesis, takes place within a single, shared cytoplasm. These events, all of which require substantial coordination, include the processes of nuclear replication and migration, formation of the germline progenitor cells, cyclic rearrangements of the embryonic cortex during late syncytial stages, and simultaneous formation of the cellular blastoderm, through the process of cellularization (Fig. 1).

The microtubule- and microfilament-based cytoskeletons both play major roles in mediating the dynamic events which take place in the interior and at

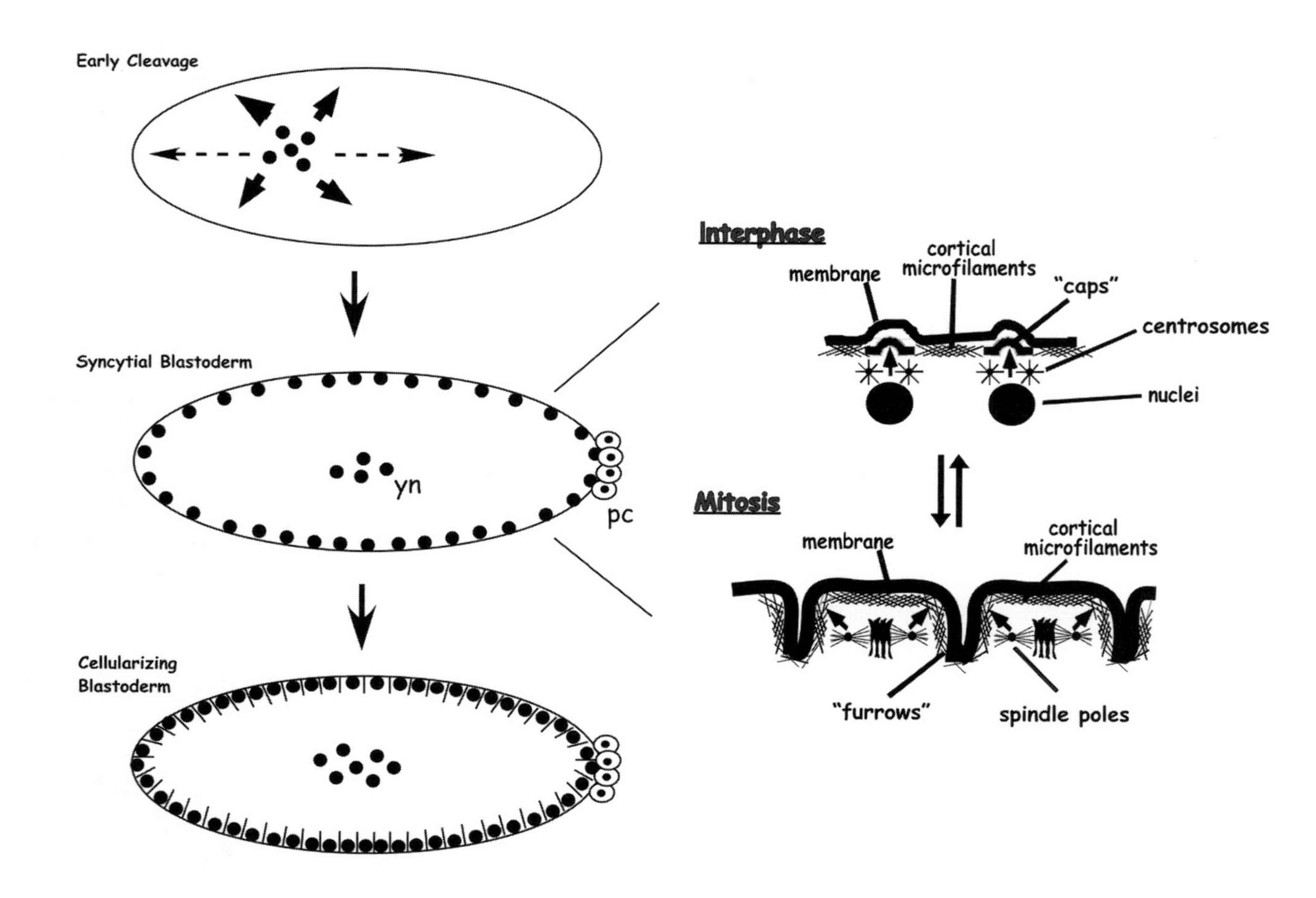
Early Cleavage
Syncytial Blastoderm
yn
pc
Cellularizing Blastoderm
Interphase
membrane
cortical microfilaments
"caps"
centrosomes
nuclei
Mitosis
membrane
cortical microfilaments
"furrows"
spindle poles

FIG. 1. Schematic representation of early *Drosophila* embryogenesis. Left panels depict progression through syncytial development. (Top) During initial stages, nuclei divide in unison internally and migrate, first along the anterior–posterior axis of the egg (dashed arrows) and then towards the cortex (solid arrows). (Middle) Cleavage cycles 10–13, the final four simultaneous divisions of the syncytial nuclei, take place in a cortical monolayer. Germline progenitor 'pole cells' (pc) form separately at the posterior end of the embryo. A small group of 'yolk nuclei' (yn) remains in the interior. (Bottom) Cellularization, during which the somatic nuclei are simultaneously enclosed in individual cells, commences at the onset of interphase of cycle 14. The illustration on the right details the structural transitions at the cortex during syncytial blastoderm stages. As described in the text, MTOCs direct (arrows) the redistribution of cortical microfilaments and structure of the surface membrane.

the cortex of the syncytial and cellularizing *Drosophila* blastoderm (Schejter & Wieschaus 1993, Sullivan & Theurkauf 1995). The early *Drosophila* embryo has proven a superb source for characterization of cytoskeleton-related proteins by biochemical means (Miller & Kellogg 1994). However, it is the amenability of this system to genetic analysis that presents a unique opportunity to explore cytoskeletal function. The cellular process at the focus of this chapter is the involvement of the cytoskeleton in establishing order within the spatial array of syncytial nuclei. I will both briefly review the published record and present new data relating to the identification and characterization, via a genetic approach, of cytoskeletal elements that influence this key aspect of cytoplasmic organization in the early embryo.

To appreciate the experimental basis for the genetic approach to early *Drosophila* embryogenesis, it is crucial to realize that the zygotic genome begins to contribute in a major way to morphogenetic events, only following completion of the syncytial period of synchronous nuclear cleavage cycles and the onset of cellularization (Merrill et al 1988). Early embryonic stages are thus governed by maternally derived gene products, deposited in the developing egg during oogenesis. As a result, mutations that affect morphogenesis of the early embryo are almost strictly maternal in origin. Given this reliance on maternally supplied factors, mutations affecting early cytoskeletal functions have been traditionally recovered among female-sterile alleles. However, mutations in genes which in addition to their roles during early embryogenesis are also required zygotically during later developmental stages, cannot be identified through such an approach. Since many elements of the cytoskeleton are expected to play essential roles in a variety of processes throughout development, this limitation helps to explain the relative paucity of mutations in early cytoskeletal functions. The classical method employed to circumvent this difficulty and study the maternal requirement for zygotically essential genes involves the production of a mosaic germline in females heterozygous for a lethal mutation. The introduction of the yeast FRT-FLP recombination system into *Drosophila* has enabled researchers to turn this challenging technique into a generally applicable tool (Chou & Perrimon 1996), leading to a significant expansion of the repertoire of known mutations affecting early embryonic events. The studies described below make use of both female sterile alleles and germline clone analysis of mutant alleles in essential genes.

Nuclear migration

Movement and positioning of nuclei within the common embryonic cytoplasm can be divided into two major processes:

- Ordered, stepwise migration of nuclei during the early cleavage cycles, initially deep within the egg and subsequently from the interior of the egg to the cortex;
- Establishment of a crowded, yet well-organized monolayer of dividing nuclei in the cortical cytoplasm, during the final four division cycles, prior to cellularization.

Nuclear migration has been explored primarily through descriptive studies, and the effects of cytoskeletal disruption on the process have been monitored as well. Mechanistic explanations for the different phases, invoking key roles for microfilament and microtubule-based structures have been proposed on the basis of these studies (Baker et al 1993, von Dassow & Schubiger 1994, Foe et al 2000). Genetic analysis of nuclear migration has lagged behind the descriptive and cell-biological studies. The potential usefulness of this approach is demonstrated, for example, by the observation that migration is abnormal when levels of the gene *sqh*, which encodes the *Drosophila* cytoplasmic myosin regulatory light-chain homologue, are reduced (Wheatley et al 1995). This has been complemented by a more recent study demonstrating dynamic changes in the cortical levels of the *sqh* gene product, leading in turn to a model invoking a key role for cortical acto-myosin contractions in nuclear migration (Royou et al 2002). Several additional mutations that display a migratory phenotype have been described (e.g. Hatanaka & Okada 1991), but their molecular nature has yet to be ascertained. Such an advance is bound to contribute substantially to detailed, molecular level description of nuclear motion within the embryonic cytoplasm.

Cortical microfilament structures

The major process to be discussed here is the restructuring of embryonic cortical microfilaments following migration of the nuclei from the interior. Arrival of the migrating nuclei at the cortex of the embryo leads to extensive reorganization of the embryonic surface and the underlying cytoskeleton. The restructuring is periodic in nature, as it follows the nuclear division cycles. This is a particularly well-studied process, analysed repeatedly in both fixed and living embryo preparations (Warn et al 1984, Karr & Alberts 1986, Kellogg et al 1988, Foe et al 2000). With respect to the actin cytoskeleton, the previously uniform cortical layer of microfilaments responds to the nuclei by alternating between concentration in supra-nuclear 'caps' during interphase and lining of transient membrane invaginations or 'furrows' during mitosis (Fig. 1). These structures act to limit nuclear motion within the cortical cytoplasm and place barriers between neighbouring nuclei, in lieu of permanent cell membranes.

While descriptive studies have provided a detailed picture of the dynamic behaviour of the major cytoskeletal elements, the genetic approach has singled out specific functional components allowing, in turn, for greater insight and formulation of ideas regarding the mechanistic basis of cyclic cortical restructuring. Several prominent examples that help illustrate this point include:

- *nuclear fallout*—involvement of this component of recycling endosomes in the process of mitotic furrow formation suggests a functional connection between recruitment of membrane vesicles and redistribution of microfilaments, and/or of factors that promote actin polymerization (Riggs et al 2003).
- *Abl*—the *Drosophila* homologue of the Abelson tyrosine kinase is required for proper formation of cortical microfilament structures, possibly acting to spatially restrict the actin polymerization activity of Enabled/VASP (Grevengoed et al 2003).
- *Sqh*—studies of myosin function during the cortical division stages, via mutations in the aforementioned regulatory light-chain homologue, address a long-standing issue, and suggest that the process of mitotic furrow invagination is not an actomyosin-based contractile event (Royou et al 2004).

Roles of Arp2/3

The Arp2/3 complex and its activators constitute a highly conserved machinery, which has been shown to promote actin polymerization in a variety of cellular settings. Characterization of Arp2/3 function during *Drosophila* development, via a genetic approach, revealed specific requirements during oogenesis and mechano-sensory bristle growth (Hudson & Cooley 2002). We (Zallen et al 2002) and others (Stevenson et al 2002) extended these studies to assess the role of the Arp2/3 system in the dynamic cortical events of the syncytial-stage embryo. Although strong mutations in Arp2/3 subunits lead to an arrest of oogenesis, weak alleles supply sufficient function for mature egg production and onset of embryogenesis following fertilization. Major conclusions regarding the roles of Arp2/3 in the early embryo include:

- The Arp2/3 complex is required for production of much of the cortical F-actin present in the early *Drosophila* embryo. Comparative quantitation of surface fluorescence in phalloidin-stained embryos revealed that Arp2/3 mutants contain about one-third of the wild-type levels of cortical F-actin (Zallen et al 2002). Since these values were obtained under circumstances where Arp2/3 function is only partially compromised, it is possible that generation of cortical microfilaments in this setting is completely dependent on proper functioning of the Arp2/3 complex.

- Specific defects in cortical microfilament structures result from abrogation of Arp2/3 function. Relatively small actin caps form at interphase of the cortical division cycles. Defective restructuring of these caps leads to a complete failure in construction of mitotic furrows, suggesting that Arp2/3 function is essential for the dynamic transition between these two microfilament-based structures (Stevenson et al 2002, Zallen et al 2002).
- Mutations in *SCAR*, the *Drosophila* homologue of the Scar/WAVE branch of Arp2/3 activators, result in highly similar early embryo phenotypes to those observed when Arp2/3 function is reduced. Both the overall reduction in cortical actin levels, and the specific defects in cap and mitotic furrow formation are observed in the SCAR mutant embryos. SCAR is thus a likely mediator of Arp2/3 function in the early embryo. In contrast, the *Drosophila* WASp homolog, representing the second major branch of Arp2/3 activators, is dispensable during syncytial stages of embryonic development (Zallen et al 2002).

MTOCs direct microfilament organization

Microtubule organizing centres (MTOCs) are thought to act as the instructive agents driving the cyclic restructuring of cortical microfilaments in early *Drosophila* embryos (Fig. 1). This major tenet in the field relies on several lines of evidence.

- There is a close spatial correspondence between MTOC position and microfilament distribution. Thus, centrosomes lie immediately beneath the interphase actin caps. The accumulation of microfilaments at the edges of the caps upon entry to mitosis, and the subsequent redistribution to invaginating furrows, furrow retraction and formation of daughter caps all follow the movement of MTOCs through the nuclear division cycle (reviewed in Rothwell & Sullivan 2000).
- We have been able to demonstrate that MTOCs are *necessary* for microfilament structure formation, via a genetic approach. Centrosomin (Cnn) is a core component of early embryonic MTOCs, and functional MTOCs are absent from embryos bearing mutations in the *cnn* gene (Megraw et al 1999, Vaizel-Ohayon & Schejter 1999). While nuclei can still divide and migrate under these circumstances, they completely fail to elicit a cortical microfilament

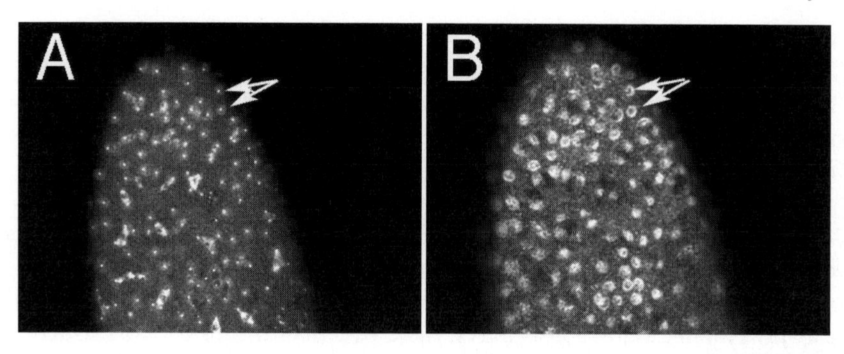

FIG. 2. Free centrosomes induce actin cap formation. (A,B) An embryo derived from a *mat(2)ea-F* mutant female (Schupbach & Wieschaus 1989), stained with anti-γ-tubulin (A) to visualize MTOCs and rhodamine-phalloidin (B) to visualize cortical microfilaments. Defects during internal cleavages in these mutant embryos can lead to generation of free centrosomes, which reach the cortex separate from the nuclei. Small actin caps consistently form above the free centrosomes (arrows).

response upon arrival at the embryonic cortex, implying a critical role for MTOCs in directing cortical microfilament restructuring (Vaizel-Ohayon & Schejter 1999).

- MTOCs are *sufficient* to generate cortical actin structures. Drug treatment, irradiation and genetic mutations have all been used to produce 'free' centrosomes in *Drosophila* embryos, which are no longer associated with nuclei. When positioned at the cortex, free centrosomes affect the formation of actin caps (e.g. Fig. 2), providing a strong demonstration of the instructive influence of MTOCs on the spatial organization of cortical microfilaments (Raff & Glover 1989, Yasuda et al 1991).

A DOCK180/CDM pathway mediates cortical events

These observations imply that a signalling pathway links MTOCs with the actin-based cytoskeleton in the cortex of the syncytial *Drosophila* embryo. What is the molecular nature of this pathway? Mutations in the strictly maternal-effect gene *sponge* (*spg*) result in phenotypes expected of a component of the MTOC–microfilament pathway. Cortical microfilaments in *spg* mutant embryos are generally oblivious to the presence of nuclei in their immediate vicinity. Apart from small cortical patches depleted of microfilaments directly above the nuclei, an unstructured, layer-like organization of F-actin is maintained during late syncytial stages (Postner et al 1992) (Fig. 3). Early, internal cleavage cycles and

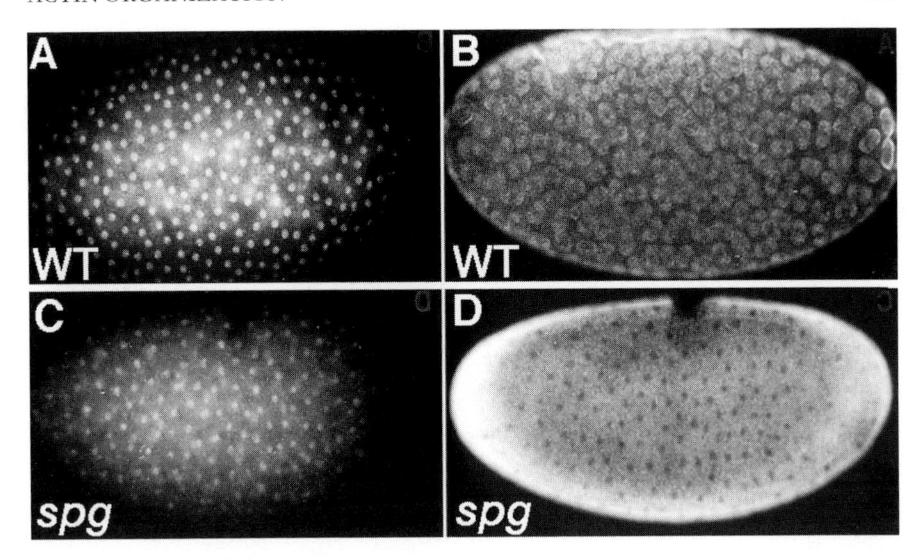

FIG. 3. The *sponge* mutant phenotype. (A, B) A wild-type (WT) cycle 10 embryo, stained with DAPI (A) and rhodamine-phalloidin (B) to reveal the spatial arrangement of cortical nuclei and microfilaments, respectively. (C, D) A *sponge* (*spg*) mutant embryo, stained with the same reagents. While robust actin caps form above the nuclei of the wild-type embryo, cortical microfilaments in the *spg* embryo retain a thick, layer-like appearance, interrupted only by small 'holes' above the nuclei.

nuclear migration to the cortex proceed normally in *spg* embryos, and so the cortical phenotypes are unlikely to be a secondary consequence of division cycle defects. Furthermore, while the majority of described mutations that disrupt cortical organization affect specific aspects of the microfilament restructuring process, the *spg* phenotype is maintained throughout the cortical cycles, regardless of division cycle phase. These phenotypes suggest that *spg* plays a key and specific role in mediating an instructive signal between functional MTOCs and the cortical microfilament layer.

We have determined that *spg* encodes a *Drosophila* homologue of the conserved DOCK180 protein family. DOCK180-like proteins are also referred to by the acronym CDM, derived from the first initial of the three founding members of the family, CED-5 (a *Caenorhabditis elegans* apoptosis and cell-migration factor) (Wu & Horvitz 1998), mammalian DOCK180 (Hasegawa et al 1996), and a second *Drosophila* homologue, Myoblast City (MBC), which is required during embryonic myogenesis and other developmental processes (Erickson et al 1997). As is apparent even from this short list, CDM proteins contribute to morphogenetic events in a wide variety of cellular and developmental settings. At the molecular level, CDM proteins are thought to act as part of a complex, downstream of cell-surface receptors, and upstream of the actin-based cytoskeleton. Additional components of the complex include the SH2-SH3 adapter protein Crk (Hasegawa et al 1996), and Elmo/CED-12, which functions together with CDM proteins as an unconventional guanine nucleotide exchange factor (GEF) for Rac GTPases (Brugnera et al 2002). CDM proteins define a subset of a larger superfamily of DOCK180-like proteins, which may act as GEFs for a wide spectrum of small GTPases (Cote & Vuori 2002).

Conclusions

The actin-based cytoskeleton is a major contributor to the cellular mechanisms employed in organizing the cytoplasm of the early, syncytial *Drosophila* embryo. Genetic studies have identified key molecular components that mediate microfilament function in this unique developmental setting. Our work in this context has revealed roles for two prominent and conserved protein complexes, known to link signalling pathways with microfilament dynamics. The Arp2/3 system acts to establish the early embryonic cortical microfilament array and in formation of specific actin-based structures. Sponge, a *Drosophila* DOCK180/CDM homologue, appears to mediate the instructive signalling through which MTOCs direct cortical microfilament behaviour. Future studies should focus on the detailed mechanisms employed within these molecular frameworks.

References

Baker J, Theurkauf WE, Schubiger G 1993 Dynamic changes in microtubule configuration correlate with nuclear migration in the preblastoderm Drosophila embryo. J Cell Biol 122:113–121

Brugnera E, Haney L, Grimsley C et al 2002 Unconventional Rac-GEF activity is mediated through the Dock180-ELMO complex. Nat Cell Biol 4:574–582

Chou TB, Perrimon N 1996 The autosomal FLP-DFS technique for generating germline mosaics in Drosophila melanogaster. Genetics 144:1673–1679

Cote JF, Vuori K 2002 Identification of an evolutionarily conserved superfamily of DOCK180-related proteins with guanine nucleotide exchange activity. J Cell Sci 115:4901–4913

Erickson MR, Galletta BJ, Abmayr SM 1997 Drosophila myoblast city encodes a conserved protein that is essential for myoblast fusion, dorsal closure, and cytoskeletal organization. J Cell Biol 138:589–603

Foe VE, Field CM, Odell GM 2000 Microtubules and mitotic cycle phase modulate spatiotemporal distributions of F-actin and myosin II in Drosophila syncytial blastoderm embryos. Development 127:1767–1787

Grevengoed EE, Fox DT, Gates J, Peifer M 2003 Balancing different types of actin polymerization at distinct sites: roles for Abelson kinase and Enabled. J Cell Biol 163: 1267–1279

Hasegawa H, Kiyokawa E, Tanaka S et al 1996 DOCK180, a major CRK-binding protein, alters cell morphology upon translocation to the cell membrane. Mol Cell Biol 16: 1770–1776

Hatanaka K, Okada M 1991 Retarded nuclear migration in Drosophila embryos with aberrant F-actin reorganization caused by maternal mutations and by cytochalasin treatment. Development 111:909–920

Hudson AM, Cooley L 2002 A subset of dynamic actin rearrangements in Drosophila requires the Arp2/3 complex. J Cell Biol 156:677–687

Karr TL, Alberts BM 1986 Organization of the cytoskeleton in early Drosophila embryos. J Cell Biol 102:1494–1509

Kellogg DR, Mitchison TJ, Alberts BM 1988 Behaviour of microtubules and actin filaments in living Drosophila embryos. Development 103:675–686

Megraw TL, Li K, Kao LR, Kaufman TC 1999 The centrosomin protein is required for centrosome assembly and function during cleavage in Drosophila. Development 126: 2829–2839

Merrill PT, Sweeton D, Wieschaus E 1988 Requirements for autosomal gene activity during precellular stages of Drosophila melanogaster. Development 104:495–509

Miller KG, Kellogg DR 1994 Isolation of cytoskeletal proteins from Drosophila. Methods Cell Biol 44:259–277

Postner MA, Miller KG, Wieschaus EF 1992 Maternal effect mutations of the sponge locus affect actin cytoskeletal rearrangements in Drosophila melanogaster embryos. J Cell Biol 119:1205–1218

Raff JW, Glover DM 1989 Centrosomes, and not nuclei, initiate pole cell formation in Drosophila embryos. Cell 57:611–619

Riggs B, Rothwell W, Mische S et al 2003 Actin cytoskeleton remodeling during early Drosophila furrow formation requires recycling endosomal components Nuclear-fallout and Rab11. J Cell Biol 163:143–154

Rothwell WF, Sullivan W 2000 The centrosome in early Drosophila embryogenesis. Curr Top Dev Biol 49:409–447

Royou A, Sullivan W, Karess R 2002 Cortical recruitment of nonmuscle myosin II in early syncytial Drosophila embryos: its role in nuclear axial expansion and its regulation by Cdc2 activity. J Cell Biol 158:127–137

Royou A, Field C, Sisson JC, Sullivan W, Karess R 2004 Reassessing the role and dynamics of nonmuscle myosin II during furrow formation in early Drosophila embryos. Mol Biol Cell 15:838–850

Schejter ED, Wieschaus E 1993 Functional elements of the cytoskeleton in the early Drosophila embryo. Annu Rev Cell Biol 9:67–99

Schupbach T, Wieschaus E 1989 Female sterile mutations on the second chromosome of Drosophila melanogaster. I. Maternal effect mutations. Genetics 121:101–117

Stevenson V, Hudson A, Cooley L, Theurkauf WE 2002 Arp2/3-dependent pseudocleavage [correction of psuedocleavage] furrow assembly in syncytial Drosophila embryos. Curr Biol 12:705–711

Sullivan W, Theurkauf WE 1995 The cytoskeleton and morphogenesis of the early Drosophila embryo. Curr Opin Cell Biol 7:18–22

Vaizel-Ohayon D, Schejter ED 1999 Mutations in centrosomin reveal requirements for centrosomal function during early Drosophila embryogenesis. Curr Biol 9:889–898

von Dassow G, Schubiger G 1994 How an actin network might cause fountain streaming and nuclear migration in the syncytial Drosophila embryo. J Cell Biol 127:1637–1653

Warn RM, Magrath R, Webb S 1984 Distribution of F-actin during cleavage of the Drosophila syncytial blastoderm. J Cell Biol 98:156–162

Wheatley S, Kulkarni S, Karess R 1995 Drosophila nonmuscle myosin II is required for rapid cytoplasmic transport during oogenesis and for axial nuclear migration in early embryos. Development 121:1937–1946

Wu YC, Horvitz HR 1998 C. elegans phagocytosis and cell-migration protein CED-5 is similar to human DOCK180. Nature 392:501–504

Yasuda GK, Baker J, Schubiger G 1991 Independent roles of centrosomes and DNA in organizing the Drosophila cytoskeleton. Development 111:379–391

Zallen JA, Cohen Y, Hudson AM, Cooley L, Wieschaus E, Schejter ED 2002 SCAR is a primary regulator of Arp2/3-dependent morphological events in Drosophila. J Cell Biol 156:689–701

DISCUSSION

Drubin: You showed that before the nuclei go out to the periphery there is already a dense network present, which you said was rather featureless. Do you view this process as a reorganization of the existing actin or as a new process?

Schejter: There is cycling of cortical actin levels from the time that the nuclei reach the periphery. This is something we and others have observed (Foe et al 2000, Zallen et al 2002). We have some evidence of global microfilament depolymerization as nuclei reach the cortex, followed by active polymerization that is contributing to formation of these various structures. I can't be much more specific about this.

Drubin: Does the localization of Scar provide any clues as to what is happening?

Schejter: The antibody against Scar wasn't very informative in the early embryos. Arp2/3 does appear at the edges of the caps, presumably at the site where it is influencing the transition between caps and furrows.

Harden: How does the strength of the phenotype of the Rac mutants compare with the *sponge* mutants? I thought that many *rac* triple mutant embryos could progress well into embryogenesis.

Schejter: If you take out all three from the female germline, no eggs are laid and you don't get an early embryo. The *rac1 rac2* double mutant we used is thus not a complete *rac* loss-of-function circumstance. Most of these embryos display some sort of cortical actin phenotype, but it is variable, however, and many are able to overcome whatever defects they show in the syncytium and go on further.

Harden: Similarly, with Cdc42 you can't address this.

Schejter: We don't get anything with Cdc42 germline clones—no eggs are laid.

Harden: Your work shows that *Drosophila* Wasp does not seem to be a major route of action for Cdc42 regulation. Can Scar/Wave be regulated by Cdc42 as well as Rac? There is only one Scar and one Wasp in *Drosophila*.

Schejter: I don't know of any obvious place where Cdc42 phenotypes and Scar phenotypes intersect.

Harden: One other comment: we noticed that Pak is associated with those transient metaphase furrows (Harden et al 1996).

Gundersen: You have a nice candidate for something that is regulating the actin, but what is the connection to the centrosome?

Schejter: That's an important question. We don't have any good genetic clue as to how that is working.

Gundersen: Is there evidence of interaction of any of the DOCKs with microtubules or centrosomal proteins?

Schejter: Not that I am aware of. There is no close to obvious explanation as to how this might work out, i.e. how a DOCK180-type complex is influenced by MTOCs. Our approach to it would be to search for additional, informative genetic mutations.

Vallee: Do you need centrosomes at all? If you put in a bead which nucleated microtubules, would this induce furrow formation?

Schejter: In the *centrosomin* mutant there are no centrosomes and there are no caps or furrows.

Vallee: There are also no astral microtubules. The issue is whether it is the astral microtubules or the centrosome. You could artificially create an aster. It is likely that it is a signal emanating from plus end transport along the microtubules rather than anything to do with the centrosome itself.

Schejter: I guess I was using 'centrosome' in a rather loose sense. It could be acting primarily as a source for a polarized microtubule array, as you suggest. To the best of my knowledge, experiments to directly address this idea have not been performed.

Chang: Can you inhibit microtubules and still get the centrosomes to form caps?

Schejter: Some studies addressing this issue have been reported. In these experiments microtubule organization in the cortex of syncytial embryos was disrupted by cold treatment (Callaini et al 1991) or by treatment with colchicine (Stevenson et al 2001), but this did not prevent formation of actin-based structures.

Chang: Are microtubules transporting something from the centrosome to the cortex?

Schejter: The experiments to identify this have not been done.

Sheetz: There is an alternative mechanism, which wouldn't imply any enzymatic transport. It would be purely mechanical. We have seen that the stretch of isolated cell cytoskeletons activates the C3G-Crk GEF pathway for G protein exchange into the GTP form, in this case Rap1 (Tamada et al 2004). Purely mechanical stretch is sufficient to activate this enzymatic pathway. One possibility here is that the process of anaphase A occurring within this capsule of actin is sufficient to mechanically disturb this capsule and thereby mechanically activate the subsequent processes. Is that viable?

Schejter: I think it is fine. We have no strong evidence of what might be linking the two and the localization data for the DOCK180 homologue is not informative. From what you are saying it doesn't have to be localized: it could be anywhere, and simple stress could activate it.

Gundersen: Wouldn't your centrosomin mutant argue against this? There you have normal anaphase and yet you see abnormal furrowing?

Schejter: Mitosis in *centrosomin* mutant embryos is, in fact, abnormal—the spindles display a 'barrel' shape, with diffuse, unfocused poles. These microtubule arrays may not be compatible with the mechanism that was just suggested.

Balasubramanian: There is a signalling complex called the SIN (septum initiation network) pathway that is localized to the spindle pole body. It is a series of protein kinases. This pathway maintains the actomyosin ring. It resides at the spindle poles and signals to the division site, where it maintains the actomyosin ring. Cnn could be like the SIN pathway, and there could be a series of proteins, with something transported from the centrosome to the division site.

Schejter: It sounds good. What we lack in this system is a good idea of the localization of the various components that are present, including basic elements of the cytoskeleton. Despite quite a bit of work in this field over the years, a good model for how caps and furrows form and reform is still lacking.

Vallee: There is another element we are ignoring that is very special about the furrow formation. This arises from classic experiments by Ray Rappaport (1986) in which he moved asters around and also drew sand dollar eggs into a capillary tube. There are two things that are special. One is that there is a requirement for an aster, and the second is that furrow formation occurs at the site of juxtaposition of two asters. We proposed many years ago that one thing that would be special about the juxtaposition of two asters is that plus end structures (e.g. vesicles) would be enriched at the junction. Does this still fit the data?

Schejter: It probably does. Victoria Foe is a proponent of microtubules being physical tracks for material that is being carried to these sites of furrow formation, in the syncytial blastoderm (Foe et al 2000). All kinds of data suggest that we can find the type of microtubule array that would fit this model. I can't contest this model on the basis of the genetics, either. Hers is probably the most detailed morphological study that exists.

Borisy: On this issue of transporting signals, since Richard Vallee has raised the topic of Ray Rappaport's experiments with eggs, and since you have mentioned that perhaps signals could be transported along microtubules, I have to jump in and say something. Rappaport measured the rate at which signals were sent to the cortex. They were not sent from the MTOC per se; they were sent by microtubules. The rate they were sent was the rate at which microtubules grew. It was not a motor-driven process. It went with the growth of the asters. This led to the idea that there was something about the ends of the microtubules that was tickling the cortex. It relates again to the topic of capture, and the communication between the microtubules and the actin. It is more consistent with an idea of interaction with the microtubule ends and the actin.

Chang: We are trying to relate the cellularization process to cytokinesis, but I think there are some major differences between these processes.

Borisy: If it is a furrowing reaction, it may be functionally related to cytokinesis.

Chang: I beg to differ. It is probably not contractile. It is organized by Arp2/3 instead of formins.

Alberts: I beg to differ. I have the reference here (Afshar et al 2000). There are very similar phenotypes including deficiencies in the furrow invagination and cellularization—the point is that formins also probably have a role in generating multiple actin structures that could effect microtubules and vice versa at different points of the cell cycle.

Nelson: There was a paper by Bornens' group (Piel et al 2001). He used centrin-GFP and showed that during anaphase B (practically telophase) the centrosome aster appears very transiently towards the furrow and then disappears again. It is a very short-term event but it is required for cytokinesis.

Borisy: It is a very late event, at the midbody, and it is not proven to be required. This event has to do with scission of the midbody, not with furrowing. Eyal, coming back to your results on the actin organization of the cortex, you said that there is a cortical cap and then this cap spreads and makes the furrow. Did you say that the cap is produced even in the Scar or Arp mutant?

Schejter: Yes.

Borisy: So the production of the cap is independent of Scar or Arp.

Schejter: Yes, that is why I said that it is probably some kind of downstream event not immediately connected to the initial stimulation of microfilament rearrangement from this uniform layer that existed before.

Borisy: Can you speculate on how this cap is produced?

Schejter: From a genetic standpoint, not really.

Borisy: Didn't you say that you don't get the cap in the sponge mutant?

Schejter: There are only two instances where we don't see the caps. Most of the mutations that have been worked on display a defect during this transition between a cap and a furrow. There isn't the converse situation where you get a furrow and don't get a cap. All the mutants at least have a cap and then sometimes don't make a furrow. The two mutants that don't make either are centrosomin and sponge. This is not a lot to go on.

Borisy: Do you want to pick up on Art's suggestion that formins also contribute to this process? You could have an Arp-dependent process and a formin-dependent process.

Alberts: I am a firm believer in the collaboration of the two activities. It is clear that there are overlapping actin remodelling activities associated with cleavage furrow formation and these types of processes. I am sitting here wondering about whether the DOCK180-like molecule could be regulating this. No one really knows which GTPase regulates fly diaphanous.

Schejter: We haven't really looked at diaphanous at all. It has an effect but I haven't looked to see how specific this effect is. The paper you cited from Wasserman's group again suggests a role during the cap to furrow transition. It also has an effect on the formation of permanent cell membranes during cellularization, but the effect is a subtle one.

Borisy: Since you have brought mDia back into the picture, it reminds me that we didn't hear from Greg Gundersen what mDia is doing in his microtubule capture scheme.

Gundersen: It is interacting with EB1 and APC.

Borisy: It is 'interacting'. Is that the best we can do?

Gundersen: That's as far as I'm willing to go. We think the complex it forms with EB1 and APC may be important for stabilizing microtubules and we are currently testing this idea. Eyal, in your sponge mutant you showed the spindle up near the cortex, and there was a hole in the actin. Does this mean that the spindle has gone up to the cortex and cleared out the actin, but it hasn't reassembled it?

Schejter: The sponge phenotype differs somewhat from the centrosomin mutant phenotype, where there is no signal. In that case the actin remains as it was before. There is a hole above the nucleus in sponge, which stays there throughout the division cycle. It gets a little wider at mitosis and a little smaller in interphase. We don't know the basis for this. It was also taken as an early indication that Sponge was somewhere downstream of the original signal. This hole is not a common feature of the wild-type process. Either you have trapped it in some transient period where it can't go on and then see a hole, or there is some other explanation. However, it is difficult to ascertain whether cycles of polymerization and depolymerization persist in the mutant: this might be part of what is going on. The actin could begin to depolymerize and then not enough actin is made to do anything else.

References

Afshar K, Stuart B, Wasserman SA 2000 Functional analysis of the Drosophila diaphanous FH protein in early embryonic development. Development 127:1887–1897

Callaini G, Dallai R, Riparbelli MG 1991 Microfilament distribution in cold-treated Drosophila embryos. Exp Cell Res 194:316–321

Foe VE, Field CM, Odell GM 2000 Microtubules and mitotic cycle phase modulate spatiotemporal distributions of F-actin and myosin II in Drosophila syncytial blastoderm embryos. Development 127:1767–1787

Harden N, Lee J, Loh HY et al 1996 A Drosophila homolog of the Rac- and Cdc42-activated serine/threonine kinase PAK is a potential focal adhesion and focal complex protein that colocalizes with dynamic actin structures. Mol Cell Biol 16:1896–1908

Piel M, Nordberg J, Euteneuer U, Bornens M 2001 Centrosome-dependent exit of cytokinesis in animal cells. Science 291:1550–1553

Rappaport R 1986 Establishment of the mechanism of cytokinesis in animal cells. Int Rev Cytol 105:245–281

Stevenson VA, Kramer J, Therkauf WE 2001 Centrosomes and the Scrambled protein coordinate microtubule-independent actin reorganization. Nat Cell Biol 3:68–75

Tamada M, Sheetz MP, Sawada Y 2004 Activation of a signaling cascade by cytoskeleton stretch. Dev Cell 7:709–718

Zallen JA, Cohen Y, Hudson AM, Cooley L, Wieschaus E, Schejter ED 2002 SCAR is a primary regulator of Arp2/3-dependent morphological events in Drosophila. J Cell Biol 156:689–701

Epithelial cell shape and Rho small GTPases

Yasuyuki Fujita* and Vania Braga†[1]

**MRC Laboratory for Cell and Molecular Biology, Cell Biology Unit and Department of Biology, University College London, London, and †Cell and Molecular Biology Section, Division of Biomedical Science, Faculty of Medicine, Imperial College London, London, UK*

Abstract. In epithelial cells, morphology is intrinsically related to function. Typically, polarization (i.e. acquisition of a cuboidal cell shape) must occur prior to terminal differentiation and functionality of epithelial sheets. Extensive work has been performed to understand the initial steps that drive cell–cell contact assembly, a process that is essential for polarization. However, not much is known about the subsequent steps that lead to remodelling of actin cytoskeleton and concomitant cell shape changes. Here we review what is known about actin organization during epithelial polarization, and discuss current models for junction assembly, actin reorganization and signalling pathways that may contribute to the generation of a polarized epithelial morphology.

2005 Signalling networks in cell shape and motility. Wiley, Chichester (Novartis Foundation Symposium 269) p 144–158

Epithelial sheets provide a protective barrier against environmental stress and microrganisms, and play a key role in absorption and secretion. In order to exert their function effectively, epithelial cells require a polarized morphology and tight adhesion to their neighbours and underlying basement membrane. Thus an interplay between cell adhesion and cytoskeletal remodelling ensures the integrity and functionality of epithelial sheets.

Polarization is achieved when epithelial cells acquire a tall, cuboidal morphology and have defined apical and basolateral membrane domains, as assessed by differential distribution of proteins and lipids. Although the actin cytoskeleton organization in mature epithelia has been known for decades (see below), the mechanisms via which an epithelial cell polarizes is not known. We will review here the current knowledge on epithelial cytoskeleton and recent insights into how signalling pathways may contribute to epithelial morphological polarization.

[1]This paper was presented at the symposium by Vania Braga to whom correspondence should be addressed.

In mature, fully polarized epithelia, actin bundles form a circumferential ring around each cell and aligned with cell borders (Owaribe et al 1981, Zamansky et al 1991). When compared with fibroblasts, primary epithelial cells do not show prominent thick actin bundles traversing the cell body (stress fibres). Moreover, in addition to the circumferential actin organization, actin bundles also appear to terminate perpendicular to cell–cell contacts as seen by both electron microscopy and immunofluorescence (see for example Yonemura et al 1995, Vasioukhin et al 2000).

A key event in the acquisition of epithelial polarity is the establishment of cadherin dependent cell–cell contacts. Adhesive cadherin complexes interact weakly with the same type of receptors in the surface of neighbouring cells (homophilic binding). These weak interactions are reinforced and stabilized by clustering of the receptors at sites of cell–cell contacts in structures called 'puncta' and by actin recruitment to cell–cell adhesion sites (McNeill et al 1993, Adams et al 1996, 1998). Within minutes of the initial cell–cell contact, punctum appearance is visualized by immunofluorescence (Fig. 1A; McNeill et al 1993, Adams et al 1996, 1998). It is thought that cadherin clustering and actin recruitment facilitate the progression from transient to stable contacts and, subsequently, into mature junctions.

However, there is more to epithelial polarization than punctum assembly (Fig. 1B). Very little is known on the subsequent steps after puncta formation leading to global changes in cell shape. It is also unclear how the localized actin recruitment at puncta participates in the formation of the circumferential actin ring. One distinct possibility seems to be the regulation of actin dynamics in the cytoplasm by cell–cell contact formation (Gloushkankova et al 1997). Similar regulation of microtubule dynamics by junction assembly has been demonstrated previously (reviewed by Braga 2002).

Current models for actin remodelling during induction of cell–cell adhesion

At the moment, there is no comprehensive model that coordinates actin at new cell–cell contact assembly, membrane rearrangement and filament remodelling to induce morphological polarization. Available models refer to either local or global events separately. Concerning local events, two models have been proposed on how junctions are sealed together in mammalian cells. Both models have in common punctum as the unit of stable cadherin-dependent contacts and the final pattern of cadherin localization at junctions as a straight line (McNeill et al 1993, Adams et al 1996, Vasioukhin et al 2000, Ehrlich et al 2002). However, the two models differ on the cytoskeletal processes via which apposing membranes touch each other during contact formation (via lamellipodium or filopodium).

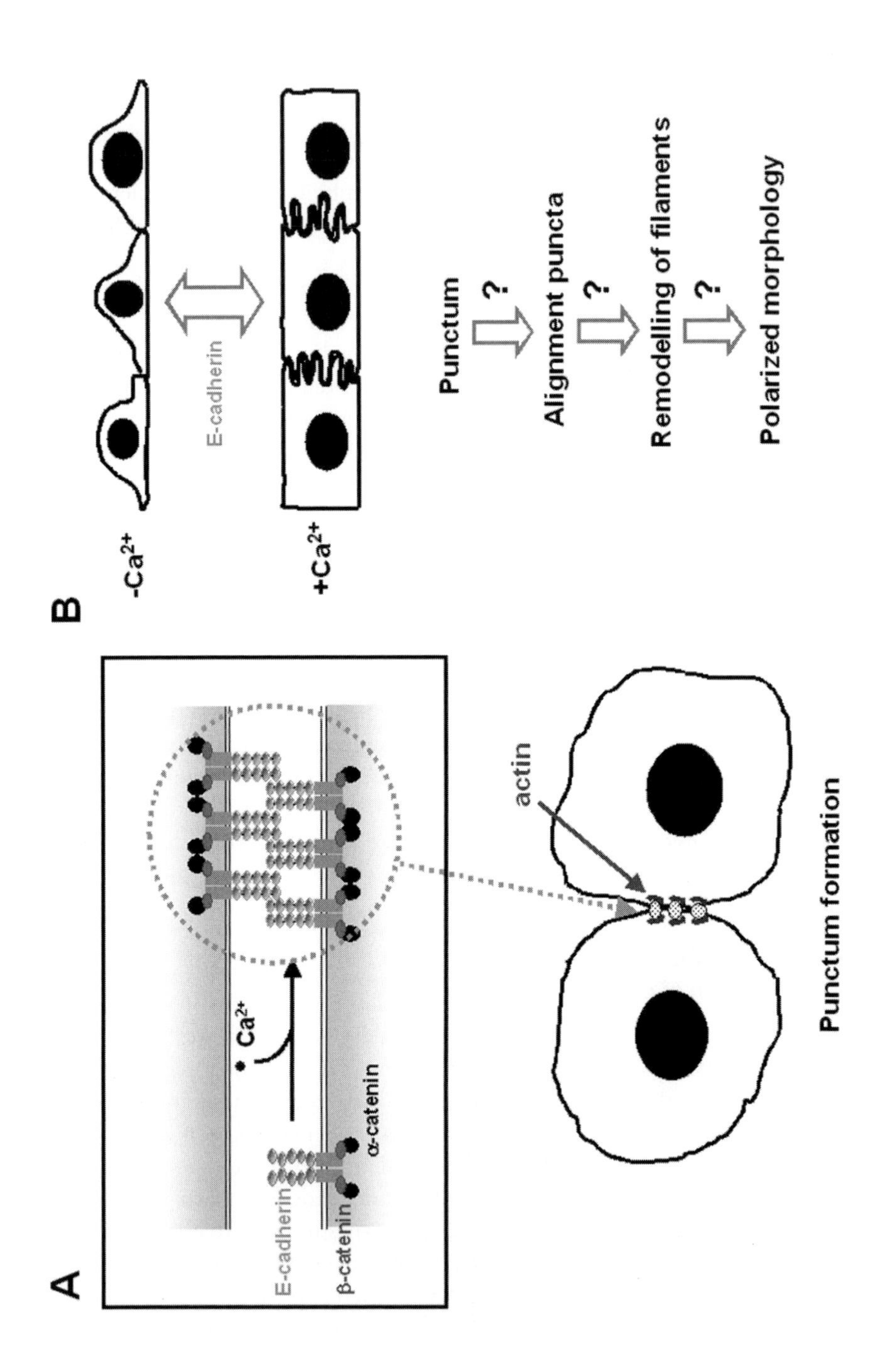
A
E-cadherin
β-catenin
α-catenin
Ca2+
actin
Punctum formation
B
-Ca2+
+Ca2+
E-cadherin
Punctum
?
Alignment puncta
?
Remodelling of filaments
?
Polarized morphology

FIG. 1. (A) E-cadherin receptors are found at the cell surface in a complex with catenins (α- and β-catenin). Cadherins become competent for adhesion in the presence of calcium ions and are then able to interact with the same type of cadherin molecules in neighbouring cells. This process leads to clustering of the receptors at sites of cell-cell contacts forming discrete puncta. Puncta are visualized at the fluorescence level as small dots of immuno-positive staining for cadherin. As soon as a punctum is formed, actin is recruited. (B) Schematic drawing of morphological changes when cell-cell contacts are induced by adding calcium ions: cells acquire a cuboidal cell shape with extensive contacts with their neighbours. Adhesion mediated by cadherins is essential for cell shape changes. Epithelial polarization involves additional steps other than punctum formation to induce a cuboidal morphology; the mechanisms and signalling molecules involved in each of the steps outlined are not known.

It is feasible that different epithelial cell types may have distinct mechanisms of sealing their membranes together. For example, electron micrographs reveal interdigitated cell–cell contacts (finger-like protrusions embedded into neighbouring cells) in newly formed and mature junctions in keratinocytes, but not in MDCK or other cell lines (Braga et al 1998, Vasioukhin et al 2000). Defining which of these models apply to a particular cell system is important as it has implications for the mechanisms via which complexes at the membrane associate with the cortical actin bundles/network junction assembly.

However, a few cautionary notes should be taken into account as the two models described above derive from sub-confluent cells that migrate to adhere to each other (Krendel et al 1999, Krendel & Bonder 1999, Vasioukhin et al 2000, Ehrlich et al 2002). First, as lamellae and filopodia have a clear role in migration, the interpretation of the contribution of these actin structures to junction assembly is complicated. Second, the finger-like protrusions observed at contacting membranes have not yet been formally characterized as filopodia (i.e. using markers such as fascin, myosin X, etc.). These finger-like protrusions may be similar to retraction fibres (Svitkina et al 2003). Consistent with that, visualization of filopodia in video of sub-confluent mammalian cells expressing GFP-actin clearly has been more difficult while lamella protruding to adhere to neighbouring cells is observed (Ehrlich et al 2002, Vaezi et al 2002). Clearly, further experiments are needed to determine the detailed organization of actin at newly clustered cadherin receptors.

In contrast to epithelial cytoskeleton, much is known regarding the actin organization in migrating cells: a dendritic network of short actin filaments present in lamellae and tight bundling of long filaments in filopodia (Bailley et al 1999, Svitkina et al 2003). However, it remains to be established how related these actin structures are to the actin present at new contact assembly sites. For example, incorporation of labelled actin in newly formed cell–cell contacts shows a distinct pattern when compared to actin incorporation into filopodia or lamella in migrating cells (pulse label experiments; Theriot & Mitchison 1991, McNeill et al 1993, Adams et al 1996, 1998, Braga et al 1997).

With respect to remodelling of bundles during epithelial polarization, two possibilities exist. The first one predicts that bundles in the cytoplasm are disassembled and reassembled closer to cell–cell contacts to form the circumferential actin ring typical of polarized epithelia. Krendel & Bonder (1999) proposed that assembly of cadherin-mediated adhesion provides a restraining force that prevents retrograde flow of membrane and actin filaments. When contacts are formed in sub-confluent cells, the typical rearward flow of filaments stops, bow-like bundles within the lamellae disassemble and re-assemble *de novo* at sites of cell–cell contacts. Thus, cell–cell contact assembly appears to modulate the typical filament dynamic observed in lamellae (rearward flow). A comparison between actin filament rearrangement during formation of cell–cell contacts in epithelia and fibroblasts revealed some similarities at the early stages of contact assembly (Yonemura et al 1995). When both cell types migrate to touch neighbouring cells, perpendicular bundles arise from cadherin-dependent adhesion sites and contact the array of circular bundles present in the cytoplasm. Dynamic studies however, have not been performed to pinpoint how this organization of perpendicular actin bundles contributes to the formation of the typical circumferential actin ring in epithelial cells. However, after contacts are established, the final organization of cytoskeleton in fibroblasts or epithelia is quite distinct: fibroblasts do not become cuboidal or have extensive cell–cell contacts with neighbours (Yonemura et al 1995, Krendel et al 1999). Differences are also observed in the shape of puncta: tangential (in epithelia) or perpendicular (in fibroblasts) to cell–cell contacts which may reflect the organization of the underlying cortical cytoskeleton (Krendel et al 1999).

A second possibility for formation of the circumferential actin ring is the remodelling of pre-existing actin filaments found in the absence of cell–cell contacts; i.e. a progressive compaction towards newly assembled junctions, until a thick band of filaments forms the circumferential ring. Further data are required to establish how the circumferential actin ring is originated after junction formation: assembly/reassembly or remodelling of pre-existing filaments.

Participation of Rho small GTPases

The Rho family of small GTPases plays an essential role in actin remodelling in many cell types. In epithelial cells, the best characterized family members, Rac, Rho and Cdc42, participate in the stabilization of cadherin receptors at junctions and in downstream events important for polarization such as differential distribution of proteins (reviewed by Van Aelst & Symons 2002). However, the precise molecular mechanisms for the functional relationship between cadherin and GTPases have not yet been dissected (Fig. 2). In particular, how GTPases function in cytoskeleton remodelling during epithelial morphological polarization is not known. Nor it is clear how Rho GTPase activities participate in the different models of junction assembly described above.

Rac is activated by cadherin-dependent adhesion in different cell types (reviewed by Braga 2002). Indeed, cadherin receptor clustering is both necessary and sufficient for Rac activation. Rac seems to participate in actin recruitment to clustered cadherin receptors in both epithelia and fibroblasts (Fig. 2; Braga et al 1997, Lambert et al 2002). Consistent with that, in MDCK cells, Rac activation leads to increased localization of both actin and cadherins at junctions (Van Aelst & Symons 2002).

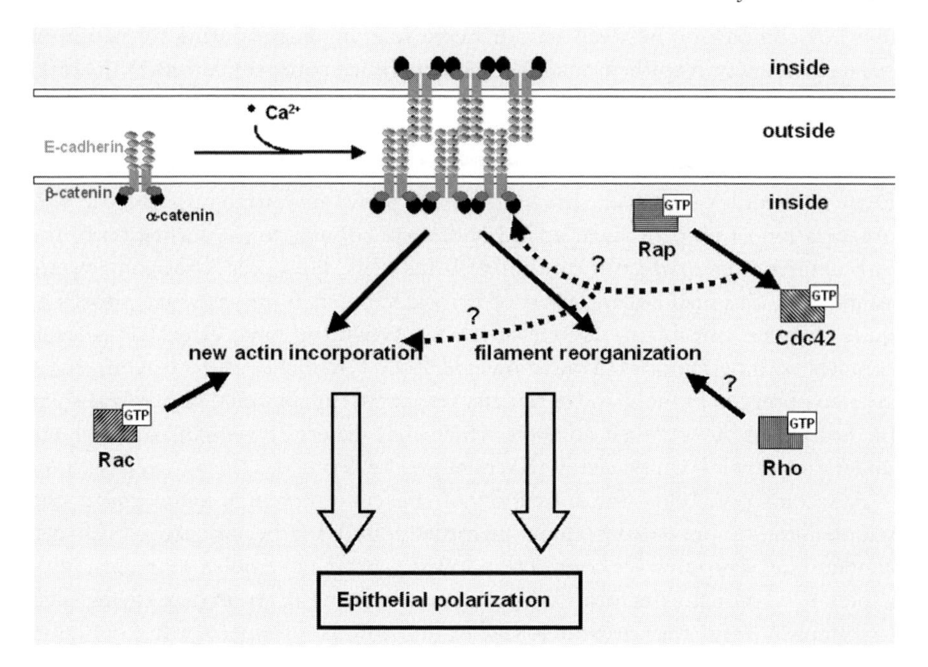

FIG. 2. Participation of small GTPases in epithelial polarization. Assembly of cadherin-dependent contacts can activate small GTPases such as Rac, Rho, Cdc42 and Rap (increased levels of interaction with GTP). The repertoire of small GTPases that are activated following junction assembly may vary depending on the cell type. Once activated, small GTPases play a role in the stabilization of cadherin receptors at junctions. The specific effects of small GTPase activation with respect to actin remodelling during polarization has only begun to be addressed. See text for more details.

Cdc42 is activated by cell–cell contact assembly in breast tumour epithelia (MCF7) but not in CHO cells expressing C-cadherin (Braga 2002). In contrast to Rac, there is yet no clear function for Cdc42 in cadherin function or cytoskeleton remodelling during polarization (Fig. 2). Instead a clearer picture of Cdc42 is emerging on its role in tight junction permeability and polarized delivery of proteins to the basolateral membrane (Braga 2002, Van Aelst & Symons 2002) Activation of Cdc42 leads to increased levels of actin at junctions (Braga 2002, Van Aelst & Symons 2002). However, whether this enhanced actin recruitment derives specifically from cadherin clustering has not yet been demonstrated. For example, many other receptors that interact with actin filaments localize at the lateral domain in epithelial cells (nectins, integrins, EGFr, etc.). Additional experiments are needed to unequivocally determine whether Cdc42 is essential for actin remodelling during epithelial polarization or indeed cadherin adhesive function (but see below).

In contrast, Rho is activated by adhesive cadherin receptors in some cell types but not others (Braga 2002). The involvement of Rho in the regulation of cadherin adhesion has been more controversial. On one hand, Rho function is necessary for the stability of cadherin receptors at junctions. However, whether Rho functions more proximal to cadherin complexes or indirectly by global remodelling of actin filaments has not been established. It is feasible that global actin filament remodelling following epithelial compaction may help to stabilize cadherin complexes at junctions, but this has not been formally demonstrated (Fig. 2). Yet, Rho activity does seem to be relevant for epithelial morphogenesis. Rho-mediated contraction of actin filaments is important for epithelial tubulogenesis, elongation of *C. elegans* or compaction in early mouse embryo (morula stage) and epithelial colonies *in vitro* (Adams et al 1998, Clayton et al 1999, Wissmann et al 1999, Wozniak et al 2003).

On the other hand, increased contractility in tumour cell lines correlates with morphological transformation (spindle cell shape). As inhibition of contractility in tumour cell lines restores cadherin-dependent adhesion and epithelial morphology, Rho activation has been suggested to negatively regulate junctions (Sahai & Marshall 2002). Yet, the two antagonistic effects of Rho in epithelia (positive or negative regulation of junctions) are not mutually exclusive. For example, upon oncogenenic transformation, Rho activity can be increased (reviewed by Lozano et al 2003). Thus, reducing the basal levels of Rho activity in tumour cell lines may bring the threshold to levels compatible with the stability of cadherin complexes at junctions. Similar effects are observed with respect to elevated Rac activity in transformed epithelia (Lozano et al 2003). Alternatively, there could be differential responses to Rho activation depending on the cell type or the mechanism of transformation (i.e. type of oncogene, increased levels of kinase or phosphatase activity, etc).

Another small GTPase that has recently been implicated in cadherin adhesion is Rap1. Rap1 is a member of the Ras subfamily and is involved in the regulation of cell proliferation, cell cycle, secretion, phagocytosis, and integrin-mediated cell adhesion (Caron 2003). Rap1 has been originally identified for its ability to suppress Ras-transformation. The initial clue that Rap1 may participate in cell–cell adhesion came from studies in *Drosophila*: loss of Rap1 function in imaginal disks specifically impairs the even distribution of DE-cadherin (*Drosophila* homologue) at cell–cell junctions (Knox & Brown 2002).

More recently, it has been shown that clustering of cadherin receptors in MCF7 epithelial cells is sufficient to activate Rap1 (Hogan et al 2004). This is achieved by a direct and transient association of C3G, a Rap1 activator, with the cadherin cytoplasmic tail. Thus puncta assembly leads to C3G recruitment and subsequent Rap1 recruitment and activation. The functional consequences are twofold (Hogan et al 2004). First, Rap1 activation is necessary for efficient assembly and/or stabilization of cadherin receptors at junctions. However, when Rap1 is blocked, a slow and delayed recruitment of cadherins to contact sites is observed, suggesting that alternative mechanisms exist. Second, Rap1 mediates activation of Cdc42 after cell–cell contact formation. Indeed, the negative effect of Rap1 inhibition on cadherin adhesion is completely rescued when Cdc42 is activated (Hogan et al 2004). How Rap1 and Cdc42 activation are entwined after junctions are formed remains to be determined (Fig. 2).

In addition to its newly discovered role in the assembly of cadherin-dependent cell–cell contacts, Rap1 may also be necessary for the maintenance of cell–cell adhesion and participate in the scattering of tumour cells during invasion. For example, deletions and mutations in the DOCK4 gene are found in tumour cell lines (Yajnik et al 2003). DOCK4 is able to specifically activate Rap1, but not other small GTPases such as Rac and Cdc42. Reconstitution of DOCK4 signalling pathways in tumour cell lines results in re-establishment of cell–cell contacts and suppression of growth in soft agar and tumour invasion. Thus, Rap1 dysfunction causes a loss of cell–cell contacts and increased invasion, contributing to malignancy.

These results are corroborated by findings that, in MDCK cells, Rap1 activation results in a phenotypic reversion of Ras-transformed (spindle cell shape) into a polarized epithelial morphology (Price et al 2004). Moreover, increased Rap1 activity suppresses scattering and loss of cell–cell contacts induced by hepatocyte growth factor (HGF). Taken together, these data suggest that Rap1 is a crucial regulator of cadherin-based cell–cell adhesions.

Based on the functional relationship between Rap1 and Cdc42, potential mechanisms via which Rap1 modulates cadherin adhesion can be suggested. For example, Cdc42 has been reported to regulate the trafficking of basolateral membrane proteins (Van Aelst & Symons 2002). One can envisage that activation of Rap1 induces activation of Cdc42, which then may facilitate directional vesicle transport of E-cadherin to newly assembled puncta, enabling neighbouring cells to contact one another. Secondly, Rap1 may provide a spatial cue for Cdc42 activation and actin recruitment at puncta, thereby stabilizing cadherin receptors (Fig. 2). A homologous system has been reported in budding yeast, where the determination of a new budding site is crucial for assembly of cortical actin patches and establishment of cell polarity (Chang & Peter 2003). In this case, deletion or activation of Rsr1, a Rap1 orthologue, causes randomization of the bud site, indicating its crucial role in bud site determination (Bender & Pringle 1989, Ruggieri et al 1992). Interestingly, it has been shown that Cdc42 is involved in bud formation and acts downstream of Rsr1 (Chant et al 1991, Park et al 2002).

Concluding remarks

How epithelial cells acquire their typical polarized cell shape has been the focus of much research in the past years. Most of the data published concern the differential distribution of membrane proteins to the apical or basolateral domain or the assembly of cadherin complex and actin recruitment at junctions. During epithelial polarization, a unifying view of how actin recruitment at puncta is coordinated with filament remodelling in the cytoplasm is long overdue. Understanding the mechanisms of formation and organization of epithelial actin cytoskeleton will provide useful insights into key regulatory molecules involved in the stabilization of cadherin-dependent adhesion.

Acknowledgements

We would like to thanks the Medical Research Council for generous support to our work.

References

Adams CL, Nelson WJ, Smith SJ 1996 Quantitative analysis of cadherin-catenin-actin reorganization during development of cell–cell adhesion. J Cell Biol 135:1899–1911
Adams CL, Chen Y-T, Smith SJ, Nelson WJ 1998 Mechanisms of epithelial cell–cell adhesion and cell compaction revealed by high-resolution tracking of E-cadherin-green fluorescent protein. J Cell Biol 142:1105–1119
Bailley M, Macaluso F, Cammer M et al 1999 Relationship between Arp2/3 complex and the barbed ends of actin filaments at the leading edge of carcinoma cells after EGF-stimulation. J Cell Biol 145:331–345
Bender A, Pringle JR 1989 Multicopy suppression of the cdc24 budding defect in yeast by CDC42 and three newly identified genes including the ras-related gene RSR1. Proc Natl Acad Sci USA 86:9976–9980
Braga VMM 2002 Cell-cell adhesion and signalling. Cur Opin Cell Biol 14:546–556
Braga VMM, Najabagheri N, Watt FM 1998 Calcium-induced intercellular adhesion of keratinocytes does not involve accumulation of β1 integrins at cell-cell contact sites and does not involve changes in the levels or phosphorylation of the catenins. Cell Adhesion Com 5:137–149
Braga VMM, Machesky LM, Hall A, Hotchin NA 1997 The small GTPases Rho and Rac are required for the establishment of cadherin-dependent cell–cell contacts. J Cell Biol 137: 1421–1431
Caron E 2003 Cellular functions of the Rap1 GTP-binding protein: a pattern emerges. J Cell Sci 116:435–440
Chang F, Peter M 2003 Yeasts make their mark. Nat Cell Biol 5:294–299
Chant J, Corrado K, Pringle JR, Herskowitz I 1991 Yeast BUD5, encoding a putative GDP-GTP exchange factor, is necessary for bud site selection and interacts with bud formation gene BEM1. Cell 65:1213–1224
Clayton L, Hall A, Johnson MH 1999 A role for the Rho-like GTPases in the polarisation of mouse eight-cell blastomeres. Dev Biol 205:322–331
Ehrlich JS, Hansen MD, Nelson WJ 2002 Spatio-temporal regulation of Rac1 localization and lamellipodia dynamics during epithelial cell-cell adhesion. Developmental Cell 3:259–270

Gloushkankova NA, Alieva NA, Krendel MF et al 1997 Cell–cell contact changes the dynamics of lamellar activity in non-transformed epitheliocytes but not in their *ras*-transformed descendants. Proc Natl Acad Sci USA 94:879–883

Hogan C, Serpente N, Cogram P et al 2004 Rap1 regulates the formation of E-cadherin-based cell-cell contacts. Mol Biol Cell 24:6690–6700

Knox AL, Brown NH 2002 Rap1 GTPase regulation of adherens junction positioning and cell adhesion. Science 295:1285–1288

Krendel M, Gloushankova NA, Bonder EM et al 1999 Myosin-dependent contractile activity of the actin cytoskeleton modulates the spatial organization of cell-cell contacts in cultured epitheliocytes. Proc Natl Acad Sci USA 96:9666–9670

Krendel MF, Bonder EM 1999 Analysis of actin filament bundle dynamics during contact formation in live epithelial cells. Cell Motil Cytoskeleton 43:296–309

Lambert M, Choquet D, Mege R-M 2002 Dynamics of ligand-induced, Rac1-dependent anchoring of cadherins to the actin cytoskeleton. J Cell Biol 157:469–479

Lozano E, Betson M, Braga VMM 2003 Tumour progression: small GTPases and loss of cell-cell adhesion. BioEssays 25:452–463

McNeill H, Ryan TA, Smith SJ, Nelson JW 1993 Spatial and temporal dissection of immediate and early events following cadherin-mediated epithelial cell adhesion. J Cell Biol 120:1217–1226

Owaribe K, Kodama R, Eguchi G 1981 Demonstration of contractility of circumferential actin bundles and its morphogenetic significance in pigmented epithelium *in vitro* and *in vivo*. J Cell Biol 90:507–514

Park HO, Kang PJ, Rachfal AW 2002 Localization of the Rsr1/Bud1 GTPase involved in selection of a proper growth site in yeast. J Biol Chem 277:26721–26724

Price LS, Hajdo-Milasinovic A, Zhao J et al 2004 Rap1 regulates E-cadherin-mediated cell-cell adhesion. J Biol Chem 279:35127–35132

Ruggieri R, Bender A, Matsui Y et al 1992 RSR1, a ras-like gene homologous to Krev-1 (smg21A/rap1A): role in the development of cell polarity and interactions with the Ras pathway in Saccharomyces cerevisiae. Mol Cell Biol 12:758–766

Sahai E, Marshall CJ 2002 ROCK and Dia have opposing effects on adherens junctions downstream of Rho. Nat Cell Biol 4:408–415

Svitkina TM, Bulanova EA, Chaga OY et al 2003 Mechanism of filopodia initiation by reorganization of a dendritic network. J Cell Biol 160:409–421

Theriot JA, Mitchison TJ 1991 Actin microfilament dynamics in locomoting cells. Nature 352:126–131

Vaezi A, Bauer C, Vasioukhin V, Fuchs E 2002 Actin cable dynamics and Rho/Rock orchestrate a polarized cytoskeletal architecture in the early steps of assembling a stratified epithelium. Dev Cell 3:367–381

Van Aelst L, Symons M 2002 Role of Rho family GTPases in epithelial morphogenesis. Genes Dev 16:1032–1054

Vasioukhin V, Bauer C, Yin M, Fuchs E 2000 Directed actin polymerization is the driving force for epithelial cell-cell adhesion. Cell 100:209–219

Wissmann A, Ingles J, Mains PE 1999 The *Caenorhabditis elegans* mel-11 myosin phosphatase regulatory subunit affects tissue contraction in the somatic gonad and the embryonic epidermis and genetically interacts with the Rac signalling pathway. Dev Biol 209: 111–127

Wozniak MA, Desai R, Solski PA, Der CJ, Keely PJ 2003 ROCK-generated contractility regulates breast epithelial cell differentiation in response to the physical properties of a three-dimensional collagen matrix. J Cell Biol 163:583–595

Yajnik V, Paulding C, Sordella R et al 2003 DOCK4, a GTPase activator, is disrupted during tumorigenesis. Cell 112:673–684

Yonemura S, Itoh M, Nagafuchi A, Tsukita S 1995 Cell-to-cell adherens junction formation and actin filament organization: similarities and differences between non-polarized fibroblasts and polarized epithelial cells. J Cell Sci 108:127–142

Zamansky GB, Nguyen U, Chou I-N 1991 An immunofluorescence study of the calcium-induced coordinated reorganization of microfilaments, keratin intermediate filaments, and microtubules in cultured human epidermal keratinocytes. J Invest Dermatol 97:985–994

DISCUSSION

Yap: We have done a bit of work with myosin 2. We suspect that the cooperativity between cadherin organization and actin may be somewhat more complex. Within 60 min of hitting cells with blebbistatin or ML7 we see a reduction in the amount of cadherin detectable in contacts. And clustering and adherens junction formation is almost undetectable. This is associated with a decrease in cell adhesion. There may be changes occurring both in contractility as well as in surface adhesion.

Braga: Definitely. The junctions are not the same in the presence of myosin inhibitors. Junctions are wavy, not as a straight line and there are fewer cadherin clusters. The point I wanted to make is that cadherin adhesion is able to form. Whether they are maintained or not is a separate question. In this sense, the bundles could help the maintenance, but clearly they are not necessary for the formation of junctional actin.

Titus: Could you explain your vision for how these filaments are organized at the junction? I can't imagine what the 3D organization might be.

Braga: It isn't known. In the literature bundles perpendicular to the junctions are described (e.g. Yonemura et al 1995, Vasioukhin et al 2000).

Titus: That's what it looks like in your pictures.

Braga: Looking at a cell from the top, by phalloidin staining we see F-actin as dots at cell–cell contacts. It seems that they coalesce. Some labs see perpendicular bundles but we don't. This is because we are looking at confluent cells which don't need to migrate to establish contacts. Imagine subconfluent cells. A single cell would have protrusions where other cells can touch via cadherins. Perpendicular bundles terminate at these contacts. You could imagine this actin structure may help to pull cells close together.

Titus: So you are looking at two different configurations of those parallel actin bundles. The actin filaments are both running in the same way, but some are thick bundles of actin and some are thin. Are these the two forms of actin you are talking about?

Braga: Yes. By electron microscopy arrays of bundles like this can be seen; there are also bundles terminating at mature junctions. However, by phalloidin staining we can't see perpendicular filaments to junctions in our system. There should be some mechanisms linking bundles to cell–cell adhesion sites, but we can't visualize this yet with our system.

Sheetz: Where are the ends of the filaments anchored?

Braga: No one has shown this. When we do actin labelling, we see it mostly incorporates as dots coincident with cadherins. We don't see any labelling of the bundles when we do incorporation assays with labelled actin.

Chang: I am confused. Are you talking about junctional actin? If you don't see perpendicular bundles, where is the actin?

Braga: What we call 'junctional actin' is the actin that is recruited to cadherins as soon as cell–cell contacts form. They start as little spots. It looks like a line when you view from above the cell. Later on, junctions mature and a very thick line of bundles appear (circumferential ring). This is what we call mature junctions.

Humphries: I thought the two membranes interdigitated.

Braga: Yes, we see that in keratinocytes. We think that when the cells are attached by the cadherins, the bundles contract, interdigitating the membranes. In MDCK cells there is not much of this interdigitation.

Lim: You showed that E-cadherin binds C3G, which activates Rap1, and Rap1 deficiency is rescued by Cdc42. Is Rap1 involved in the Cdc42 activation?

Braga: This is in MCF7 cells, not keratinocytes. If we block Rap with Rap-GAP, Cdc42 is not activated.

Lim: Have you followed them to see whether there is any pathway back to Cdc42 activation?

Kaibuchi: It is a situation similar to that in yeast. We are working on a similar system a bit. When we chelate Ca^{2+} by EGTA, we see a lot of Rap1-GTP inside the cell. Have you looked at Rap1-GTP? We see activation of Rap1 during the course of Ca^{2+} chelation.

Braga: Ca^{2+} has an effect on Rap activation.

Kaibuchi: Yes, a much higher activation occurs when Ca^{2+} is chelated, in our hands.

Braga: In the system we use, Ca^{2+} is absent all the time. What we do to exclude the effect of Ca^{2+} this is to cluster the receptors using latex beads coated with specific antibodies. If we cluster other receptors than cadherins, this does not have as much effect on Rap activation.

Kaibuchi: We can't exclude the possibility of involvement of Ca^{2+}. When we knock down Rap1a and b, the cell can't contact again.

Braga: There is some work in *Drosophila* showing that Rap could be involved in cell–cell adhesion. Perturbing Rap1 function does not disrupt cell–cell adhesion completely; junctions appear fragmented.

Kaibuchi: The cell polarity also seems to be changed in *Drosophila*. But in this case in MDCK epithelial cells, cadherin-mediated cell–cell contacts are impaired. Have you ever looked at the effect of Y compounds on the height of the cell?

Braga: Yes, and it is exactly the same as treatment with blebbstatin.

Kaibuchi: Maybe the Rho kinase system is involved.

Braga: Yes, but if Rho kinase is blocked, it doesn't affect the cadherin system.

Nelson: When you inject Rap-GAP you block adhesion and cadherin, where is the cadherin? Is it going to the cell surface?

Braga: It seems to disappear.

Kaibuchi: Cadherin appears to go to endosomes. After knockdown of Rap1, most of cadherin seems to stay in the endosomes.

Nelson: But does the protein get to the cell surface?

Kaibuchi: We did this kind of experiment. When we knock down Rap1 by siRNA, the cells become flattened, but still keep the intact cell–cell contacts. If we chelate Ca^{2+}, the cell–cell contacts are perturbed in the knockdown cells more rapidly than in the control cells. When Ca^{2+} is added back to the medium, the control cells form cell–cell contacts soon, whereas the Rap1-knockdown cells do not from cell–cell contacts again and most of cadherin appears to stay in the endosome-like structures.

Yap: Have you tracked it by metabolic labelling? Is it getting to the surface and then leaving?

Kaibuchi: No, not yet. But, we have found that cadherin appears at the cell–cell contact sites again soon after addition of Ca^{2+}, whereas most of the cadherin appears to stay in the endosome-like structures in the knockdown cells.

Borisy: I'd like to return to your description of the actin organization of the region of cell–cell contact. Elaine Fuchs describes the protrusions from cells as filopodia.

Nelson: Let's be clear on this. The *Cell* paper says that they are filopodia (Vasioukhin et al 2001). There is a *Developmental Cell* paper (Vaezi et al 2002) which shows clearly that it is lamellipodia and the cells then retract to leave filopodia behind.

Yap: In the caption to one of the figures in the latter paper they acknowledge retraction occurs.

Braga: It hasn't been characterized molecularly.

Nelson: The mechanism may be quite similar. Even though there was an interesting difference between primary cells and cell lines, I think the mechanisms are rather similar that both primary keratinocytes and other cells adapt mechanisms of actin dynamics to initiate cell–cell adhesion. The difference with Vania Braga's work is that she uses a different approach to look at adhesion than the Fuchs lab. The Fuchs lab use lower density cultures and the cells have to find each other. There is an issue here about cell migration, and the perpendicular protrusions of actin filaments that the Fuchs lab sees may have something to do with integrins and cell migration, something you exclude from your studies.

Braga: We have actually done the same experiments using subconfluent cells. The time courses are distinct. In Elaine Fuchs' paper they see actin formation at junctions in 3 h; in our system, actin is incorporated at junctions within minutes. By using subconfluent cultures we see the same things happening but the processes are delayed.

Borisy: In that first paper from the Fuchs lab, they specifically stated that directed actin polymerization is driving cell–cell contact.

Nelson: In the *Developmental Cell* paper this is 'refined'.

Braga: Their work is based on a published paper on *C. elegans* ventral closure. This is migration of cells; it is a different process.

Borisy: It is not disputed that the cell–cell contacts in dorsal closure are mediated by filopodial interactions, is it?

Braga: No.

Nelson: The other difference between the *Cell* paper and *Developmental Cell* paper is that the *Cell* paper is based on fluorescence imaging of fixed time points, whereas the other was using live cell imaging. The dorsal closure microscopy was elegant and showed filopodia and lamellipodia very nicely.

Harden: Live microscopy imaging shows the cells seeking each other out in dorsal closure of the *Drosophila* embryo as the two sheets come up to each other (Jacinto et al 2000).

Nelson: There are lamellipodia there as well.

Harden: They also showed electron microscopy images of the final closed dorsal surface, demonstrating interdigitation (Jacinto et al 2000).

References

Jacinto A, Wood W, Balayo T, Turmaine M, Martinez-Arias A, Martin P 2000 Dynamic actin-based epithelial adhesion and cell matching during *Drosophila* dorsal closure. Curr Biol 10:1420–1426

Vaezi A, Bauer C, Vasioukhin V, Fuchs E 2002 Actin cable dynamics and Rho/Rock orchestrate a polarized cytoskeletal architecture in the early steps of assembling a stratified epithelium. Dev Cell 3:367–381

Vasioukhin V, Bauer C, Yin M, Fuchs E 2000 Directed actin polymerization is the driving force for epithelial cell–cell adhesion. Cell 100:209–219

Vasioukhin V, Bauer C, Degenstein L, Wise B, Fuchs E 2001 Hyperproliferation and defects in epithelial polarity upon conditional ablation of alpha-catenin in skin. Cell 104:605–617

Yonemura S, Itoh M, Nagafuchi A, Tsukita S 1995 Cell-to-cell adherens junction formation and actin filament organization: similarities and differences between non-polarized fibroblasts and polarized epithelial cells. J Cell Sci 108:127–142

Interaction of cadherin with the actin cytoskeleton

W. James Nelson, Frauke Drees and Soichiro Yamada

Department of Molecular and Cellular Physiology, Beckman Center for Molecular and Genetic Medicine, 279 Campus Drive, Stanford University School of Medicine, Stanford, CA 94305-5435, USA

Abstract. Cadherins regulate cell–cell adhesion throughout embryonic development and in the adult organism, and defects in cadherin expression and function are characteristic of many disease states including cancer. Although extracellular binding between cadherins specifies adhesion between cells, the strength of the interaction is thought to be regulated by cadherin clustering through reorganization of the actin cytoskeleton. Protein–protein interactions have been described that could link cadherins either directly or indirectly to the actin cytoskeleton. Here, we describe these protein interactions, and examine critically the evidence that they link cadherins to the actin cytoskeleton.

2005 Signalling networks in cell shape and motility. Wiley, Chichester (Novartis Foundation Symposium 269) p 159–177

Cell–cell adhesion is a fundamental characteristic of multicellular organisms that specifies initial interactions between cells during formation of tissues and subsequently maintains the structural and functional integrity of those tissues (Takeichi 1995, Gumbiner 2000). Epithelial cells constitute a major cell type that contributes to the formation and organization of many tissues. Epithelial cell–cell adhesion is mediated by a variety of membrane proteins, including classical cadherins, claudins/occludin, nectin and desmosomal cadherins (Jamora & Fuchs 2002). Weak binding between extracellular domains of cadherins specifies initial adhesions between cells (Gumbiner 2000), but strong cell–cell adhesion develops during lateral clustering of cadherins. The prevailing idea is that cadherin clustering and the strengthening of cell–cell adhesion are mediated by proteins that link cadherin to the actin cytoskeleton (Jamora & Fuchs 2002); β-catenin binds to cadherin cytoplasmic domain and to α-catenin, which is linked directly or indirectly to the actin cytoskeleton. However, little is known about how these protein complexes assemble in cells, or how the cadherin/catenin complex binds and organizes the actin cytoskeleton (Fig. 1). Although much is known about actin organization at the leading edge of migratory cells (Ridley et

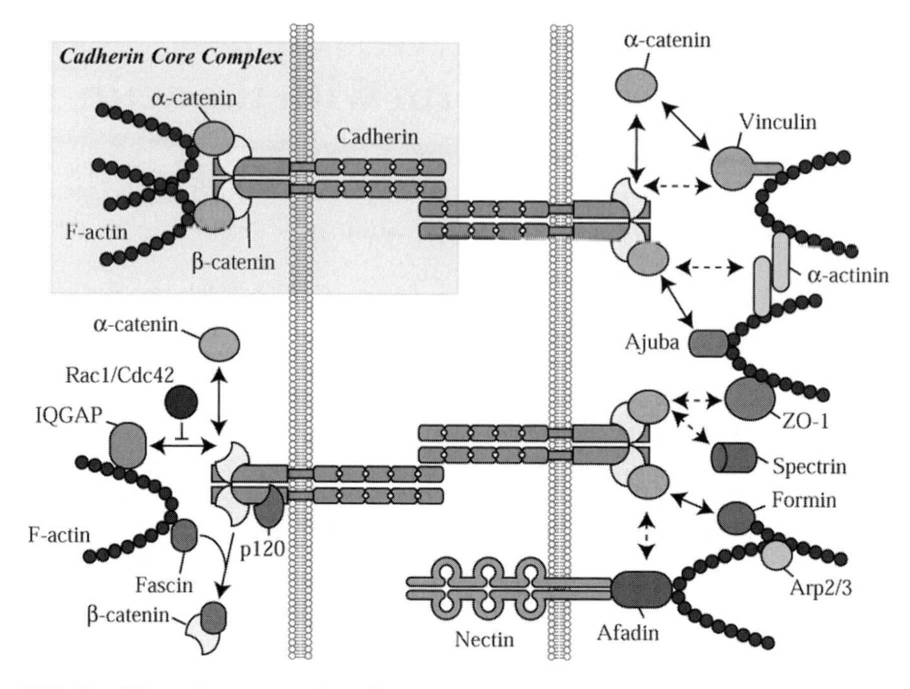

FIG. 1. Schematic representation of protein–protein interactions between the cadherin cell adhesion complex, and the actin cytoskeleton. For details, see text.

al 2003), considerably less is known about the organization of the actin cytoskeleton at sites of cell–cell contact. Here we examine critically the evidence for the role of the actin cytoskeleton in cadherin-mediated cell–cell adhesion, and the protein linkages thought to be involved.

Involvement of actin in cadherin-mediated cell–cell adhesion

Cell–cell adhesion involves dynamic interactions between the plasma membranes of opposing cells and the concomitant clustering of cadherin at sites of cell–cell contact (Adams et al 1998). Actin could be involved at several steps by regulating membrane dynamics through filament assembly and disassembly, or cadherin clustering. Evidence that cadherin linkage to the actin cytoskeleton is required for cell–cell adhesion has come from several different types of experiments. In a simple experiment, forced depolymerization of the actin cytoskeleton by addition of cytochalasin D results in the gradual disruption of cell–cell contacts. However, as actin filaments are associated with, and function in a variety of roles in many structures in cells, a global loss of the actin cytoskeleton is likely to induce a broad spectrum of changes, some of which may affect cell–cell adhesion directly or indirectly. Another approach to alter actin dynamics has been the expression of a Rac1 mutant that cannot hydrolyse GTP, a so-called dominant-negative protein that inhibits Rac1-dependent lamellipodia formation. In MDCK cells and keratinocytes, expression of this Rac1 mutant slows the kinetics of development of strong cell–cell adhesion, consistent with a role for lamellipodia in accelerating contact formation between opposing cells and, thereby, rapid recruitment of cadherin to stabilize those contacts (Braga 2000).

A more direct approach to investigating a role of actin in cell–cell adhesion is to examine cell–cell adhesion in cells lacking components thought to link cadherins to the actin cytoskeleton. Indeed, cells that express cadherin but lack α-catenin exhibit reduced cell–cell adhesion (Watabe-Uchida et al 1998). An alternative experiment is to reconstitute cadherin-mediated cell–cell adhesion in cells such as fibroblasts that lack endogenous cadherins but contain catenins. Ectopic expression of cadherins results in formation of Ca^{2+}-dependent cell–cell adhesion. Expression of cadherin containing a deletion of the cytoplasmic domain, and hence binding sites for catenins, does not enable the formation of strong Ca^{2+}-dependent cell–cell contacts, even though the extracellular domain is intact and should allow cadherin interactions between opposing cell surfaces (Sako et al 1998); some cadherin–cadherin interactions likely occur between cells, but it is assumed that they are insufficient for the compaction of cells. This result has been interpreted as a lack of actin-mediated clustering of cadherins due to deletion of the binding site for catenins and, hence, the actin cytoskeleton. This interpretation is bolstered by the finding that expression of a chimeric protein comprising E-cadherin fused to α-catenin induced cell–cell adhesion, albeit more static, in non-adherent cells, presumably through stable linkage of the chimera through the α-catenin 'domain' to the actin cytoskeleton (Sako et al 1998), suggesting that α-catenin may exhibit a more dynamic association to the cadherin core complex.

Proteins linking cadherins through α-catenin to the actin cytoskeleton

Cadherins are bound to a core complex comprising cadherin/β-catenin/α-catenin (Fig. 1), which has been isolated from epithelial cells in a stoichiometric complex following protein cross-linking (Ozawa et al 1990, Hinck et al 1994). Furthermore, high-resolution crystallographic structures of the complex between the cytoplasmic domain of E-cadherin and the armadillo repeat domain of β-catenin (Huber et al 2001), and the chimeric complex between the β-catenin/α-catenin binding regions (Pokutta & Weis 2000) have been solved. How is this core complex bound to the actin cytoskeleton? In addition to the potential direct linkage through α-catenin, many cytoplasmic actin binding proteins have been reported to interact with the cadherin core complex through α-catenin. Below is a summary of binding partners for α-catenin (Fig. 1).

Actin filaments

Rimm and colleagues (Rimm et al 1995) showed that recombinant α-catenin purified from bacteria co-sedimented with actin filaments in a stoichiometry of ~1:6, α-catenin:actin with an apparent K_d of 0.3 μM. The actin binding domains were reported to be in the N- and C-terminal of α-catenin. Electron microscopy of F-actin bound to α-catenin revealed that the actin filaments were tightly bundled compared to actin filaments in the absence of α-catenin. More recently, the C-terminal α-catenin actin binding site has been narrowed down to amino acids (aa) 671–906 (Pokutta et al 2002). These experiments have been interpreted as evidence that the cadherin core complex is bound to actin filaments directly through α-catenin. Whether α-catenin can bind to actin filaments and β-catenin simultaneously remains to be tested.

α-Actinin

Knudsen et al (1995) reported that α-actinin, an actin cross-linking protein found predominantly at focal adhesions, co-localized with N-cadherin in WI-38 fibroblasts, and the detergent-extracted pool of α-actinin could be co-immunoprecipitated with α-catenin antibodies in an α-catenin-dependent manner when extracts from large numbers of cells were used. Since no actin was co-immunoprecipitated with α-catenin/α-actinin complex, it was suggested that the binding of α-actinin to the cadherin core complex might be direct. In fact, Nieset et al (1997) reported a direct interaction of α-catenin with α-actinin by yeast two-hybrid assay, and narrowed the interacting domains of the proteins to aa 325–394 and aa 479–529 in α-catenin and α-actinin, respectively. However, the functional importance of α-actinin binding to α-catenin has not been shown in a cellular context.

Vinculin

Several groups have reported binding of vinculin to the cadherin core complex. Vinculin is an actin filament binding protein with homology to α-catenin that localizes preferentially at sites of focal adhesions. Weiss et al (1998) reported that expression of the head domain (HD) or tail domain (TD) of vinculin resulted in localization of either domain to the integrin-based focal adhesions as expected (see above), but also localization of vinculin HD to cell–cell contacts. Indeed, they showed that vinculin could be co-immunoprecipitated with α-catenin, and the cadherin core complex from PTK2 and MDBK cells, suggesting that vinculin could interact in some way with α-catenin. Using purified proteins in a dot-blot assay, Weiss et al showed that vinculin HD appeared to bind strongly to α-catenin TD, although full-length vinculin bound poorly. Furthermore, while vinculin HD did not pellet with actin filaments in a sedimentation assay, it did when α-catenin was included. BIACORE plasmon resonance spectroscopy showed that the binding affinity of vinculin HD to α-catenin was 0.2–0.4 μM, but the binding affinity of full-length vinculin was too weak to measure. The binding site for vinculin to α-catenin TD was located to aa 878–899, and α-catenin TD expressed in cells co-localized with vinculin at focal adhesions and cell–cell contacts. Watabe-Uchida et al (1998) confirmed this interaction, but in this report the vinculin binding site on α-catenin was located in the central region of α-catenin (aa 326–509). The recent solution of the crystal structure of full-length vinculin with the α-catenin vinculin binding region confirmed this domain, and calorimetry determined the K_d of this interaction to be 82 nM (Bakolitsa et al 2004). Watabe-Uchida et al (1998) took advantage of a cell line lacking α-catenin (DLD-1 cells) to express different domains of α-catenin and reported that expression of α-catenin N-terminal domain (aa 1–509) could rescue adhesion, which led to the localization of vinculin and ZO-1, but not α-actinin, to the cell–cell adhesion sites, presumably due to vinculin and ZO-1 linking to the actin cytoskeleton. Interestingly, Ozawa (1998) showed that a chimera between cadherin and the vinculin TD did not rescue cell–cell adhesion, unlike the previously reported chimera between E-cadherin and the α-catenin tail domain (Sako et al 1998).

An additional link between vinculin and the cadherin core complex has been reported through interaction with β-catenin. Hazan et al (1997) used a cell line lacking α-catenin (MDA-MB-468), and found that although they expressed E-cadherin they did not aggregate in the presence of extracellular Ca^{2+}. However, following serum starvation, there was an approximately threefold increase in the level of E-cadherin and that now ~40% of the cells aggregated.

Immunoprecipitation of either E-cadherin or β-catenin followed by Western blotting showed that vinculin was co-precipitated. The authors suggested that both α-catenin and vinculin can associate with the cadherin core complex, presumably through β-catenin, and that binding of vinculin is stimulated by increased levels of E-cadherin and facilitated by the absence of α-catenin. This was tested directly by trying to co-immunoprecipitate E-cadherin with vinculin from MDA-MB-468 cell lysates in the presence of excess recombinant α-catenin, with the result that as the concentration of α-catenin increased the amount of E-cadherin in the vinculin immunoprecipitated decreased. A straightforward conclusion from the data of Hazan et al (1997) is that vinculin binds to β-catenin with a lower affinity than that of α-catenin.

It is important to note that α-catenin and vinculin share considerable structural homology (Bakolitsa et al 2004) consistent with the possibility that they could functionally substitute for each other. However, even though there is some minor overlap in the distribution of vinculin and α-catenin at cell–cell contacts, α-catenin does not localize to integrin-based focal adhesions, indicating that there is a fundamental difference in the spatial regulation of these proteins. In this context it is noteworthy that a chimera between E-cadherin and α-catenin could induce cell–cell adhesion (Sako et al 1998), while a similar chimera between E-cadherin and vinculin TD did not (Ozawa 1998), suggesting that there have to be distinct functionalities between these two proteins. Vinculin has been shown to be auto-inhibited by its intramolecular head–tail interaction (Johnson & Craig 1995), and must undergo a conformational change upon activation by phospholipid (Gilmore & Burridge 1996). The conditions for binding studies or the nature of the truncated constructs might result in a constitutively activated protein that might be functionally different from that of full-length vinculin in the cell.

Spectrin, ZO-1

Spectrin and ZO-1 are additional structural proteins that interact with actin filaments and localize to cell–cell contacts, and have been shown to bind to specific sites on α-catenin. Pradhan et al (2001) reported that small amounts of α-catenin could be co-immunoprecipitated with spectrin, a large protein bound through ankyrin to different membrane proteins (Mohler et al 2002). They reported that α-catenin binds $\alpha 2\beta 2$ spectrin with an affinity of ∼20–100 nM, and that the binding site on α-catenin is in the head domain (aa 1–228). Earlier studies showed that E-cadherin-mediated cell–cell adhesion induced co-localization of spectrin, ankyrin and Na/K-ATPase, and it is possible that an interaction between α-catenin and spectrin could facilitate the spatial organization of Na/K-ATPase directly (McNeill et al 1990).

Itoh et al (1997) examined the distribution of ZO-1 domains in E-cadherin expressing L-cells and found that the N-terminal domain and full-length ZO-1 localized to cell–cell contacts, whereas the C-terminal domain binds actin filaments. Furthermore, the ZO-1 N-terminal domain could be immunoprecipitated with the cadherin core complex, and bound directly to recombinant α-catenin with an apparent K_d of ∼0.5 nM. Endogenous ZO-1 co-localized with the stably expressed aa 631–906 domain of α-catenin (Imamura et al 1999). In polarizing epithelial cells, ZO-1 initially co-localizes with adherens junctions but as tight junctions form ZO-1 appears to move to the tight junction (Yonemura et al 1995). While the functional significance of ZO-1/α-catenin binding in epithelial cells is poorly understood, Watabe-Uchida et al (Watabe-Uchida et al 1998), who investigated the α-catenin/vinculin interaction extensively (see above), suggested that vinculin is involved in recruiting ZO-1 to the uppermost regions of the adherens junction perhaps through α-catenin/vinculin binding.

Arp2/3 complex and formins

Although the organization of actin filaments at cell–cell contacts is not well understood, even less is known about the state of actin polymerization. Several studies have reported that purified G-actin incorporates at cell–cell contacts, suggesting that actin is polymerizing at that site (Braga et al 1997, Vasioukhin et al 2000). Consequently, several studies have sought to determine whether actin filament nucleating factors such as Arp2/3 complex and formins are also localized to cell–cell contacts. Using E-cadherin substrates and E-cadherin coated beads, Kovacs et al (2002) and Helwani et al (2004) reported that the Arp2/3 complex and its activator cortactin, respectively, co-localize with cadherin-mediated contacts, and that they can be co-immunoprecipitated with E-cadherin. However, it is unclear whether the Arp2/3 complex directly binds the cadherin core complex, or whether it is required specifically for actin assembly at cell–cell contacts. Kobielak et al (2004) showed that formins interact with α-catenin by yeast two-hybrid assay, and that the binding sites were between the FH1 and FH3 domains of formins and aa 300–500 of α-catenin. In the absence of α-catenin (an α-catenin knockout in mouse keratinocytes), neither actin nor formins localized to cell–cell contacts. However, expression of a β-catenin/formin chimera in α-catenin null keratinocytes appeared to rescue cell–cell adhesion and actin association with those sites. Other regulators of actin polymerization that have been reported to localize to cell–cell adherens junctions include Ena and VASP (Vasioukhin et al 2000), though direct interactions have not been confirmed. Together, these studies indicate that nucleating factors for actin filament assembly and their regulators are localized in the immediate vicinity of cell–cell contacts and that actin assembly is dynamic at those sites. It will be important to understand how these actin-nucleating factors are activated upon cadherin-mediated cell–cell adhesion, and the organization and dynamics of actin filaments at adhesion sites.

Additional proteins (afadin, ajuba)

Additional actin binding proteins associated with the cadherin core complex have been described at cell–cell contacts. Ajuba, a LIM domain protein, binds directly to α-catenin and is recruited to cadherin-mediated cell–cell adhesion (Marie et al 2003). Afadin from cell lysate binds to the central region of α-catenin (aa 385–651), and might thereby provide not only an additional link to the actin cytoskeleton, but also to the Nectin/afadin/ponsin adhesion system (Pokutta et al 2002).

Proteins linking β-catenin to the cytoskeleton

IQGAP is an actin binding protein activated by the small GTPases Rac1 and Cdc42. Kuroda et al (1998) found that IQGAP1 binds β-catenin, causing dissociation of the α/β-catenin complex. Activation of IQGAP by Rac1 or Cdc42 disrupted its binding to β-catenin, resulting in rebinding of α-catenin to β-catenin and, hence, functional assembly of the cadherin core complex and initiation of cell–cell adhesion. IQGAP, therefore, potentially links assembly of the cadherin core complex with actin filament organization at cell–cell contacts. β-catenin has also been identified by yeast two-hybrid as a directly interaction partner of the actin bundling protein fascin, whose binding site within β-catenin overlaps with the E-cadherin binding site, and therefore competes with E-cadherin for β-catenin binding in *in vitro* pull-down assays (Tao et al 1996). Both IQGAP1 and fascin are an example of actin binding proteins that interact with β-catenin, but have an ability to disrupt the cadherin core complexes.

Summary and future directions

It is generally assumed that the cadherin core complex is bound directly or indirectly to the actin cytoskeleton, and that this interaction is required for strong cell–cell adhesion. It remains unclear exactly how actin is organized underneath the plasma membrane at cell–cell contacts, whether it is in a state of focused assembly, disassembly or dynamic reorganization. Many proteins have been found associated with α-catenin, and in many cases the binding sites on each protein have been determined. While many interactions have been described between the cadherin core complex and actin binding proteins, it remains unclear whether they link actin filaments to the cadherin core complex, locally regulate existing actin filament organization, or locally change actin filament assembly and/or disassembly. It may be informative to attempt to reconstitute larger assemblies of proteins to test whether they interact as a complex, and whether the complex interacts with actin filaments. It will also be important to determine whether protein–protein complexes between, for example, α-catenin and vinculin, alter the binding properties of each protein or confer new binding/interacting properties with actin filaments.

Acknowledgements

Work from the Nelson laboratory is supported by a grant from the NIH (GM35227).

References

Adams CL, Chen YT, Smith SJ, Nelson WJ 1998 Mechanisms of epithelial cell-cell adhesion and cell compaction revealed by high-resolution tracking of E-cadherin-green fluorescent protein. J Cell Biol 142:1105–1119

Bakolitsa C, Cohen DM, Bankston LA et al 2004 Structural basis for vinculin activation at sites of cell adhesion. Nature 430:583–586

Braga V 2000 Epithelial cell shape: cadherins and small GTPases. Exp Cell Res 261:83–90

Braga VM, Machesky LM, Hall A, Hotchin NA 1997 The small GTPases Rho and Rac are required for the establishment of cadherin-dependent cell-cell contacts. J Cell Biol 137:1421–1431

Gilmore AP, Burridge K 1996 Regulation of vinculin binding to talin and actin by phosphatidyl-inositol-4-5-bisphosphate. Nature 381:531–535

Gumbiner BM 2000 Regulation of cadherin adhesive activity. J Cell Biol 148:399–404

Hazan RB, Kang L, Roe S, Borgen PI, Rimm DL 1997 Vinculin is associated with the E-cadherin adhesion complex. J Biol Chem 272:32 448–32 453

Helwani FM, Kovacs EM, Paterson AD et al 2004 Cortactin is necessary for E-cadherin-mediated contact formation and actin reorganization. J Cell Biol 164:899–910

Hinck L, Nathke IS, Papkoff J, Nelson WJ 1994 Dynamics of cadherin/catenin complex formation: novel protein interactions and pathways of complex assembly. J Cell Biol 125:1327–1340

Huber AH, Stewart DB, Laurents DV, Nelson WJ, Weis WI 2001 The cadherin cytoplasmic domain is unstructured in the absence of beta-catenin. A possible mechanism for regulating cadherin turnover. J Biol Chem 276:12301–12309

Imamura Y, Itoh M, Maeno Y, Tsukita S, Nagafuchi A 1999 Functional domains of alpha-catenin required for the strong state of cadherin-based cell adhesion. J Cell Biol 144:1311–1322

Itoh M, Nagafuchi A, Moroi S, Tsukita S 1997 Involvement of ZO-1 in cadherin-based cell adhesion through its direct binding to alpha catenin and actin filaments. J Cell Biol 138:181–192

Jamora C, Fuchs E 2002 Intercellular adhesion, signalling and the cytoskeleton. Nat Cell Biol 4:E101–108

Johnson RP, Craig SW 1995 F-actin binding site masked by the intramolecular association of vinculin head and tail domains. Nature 373:261–264

Knudsen KA, Soler AP, Johnson KR, Wheelock MJ 1995 Interaction of alpha-actinin with the cadherin/catenin cell-cell adhesion complex via alpha-catenin. J Cell Biol 130:67–77

Kobielak A, Pasolli HA, Fuchs E 2004 Mammalian formin-1 participates in adherens junctions and polymerization of linear actin cables. Nat Cell Biol 6:21–30

Kovacs EM, Goodwin M, Ali RG, Paterson AD, Yap AS 2002 Cadherin-directed actin assembly: E-cadherin physically associates with the Arp2/3 complex to direct actin assembly in nascent adhesive contacts. Curr Biol 12:379–382

Kuroda S, Fukata M, Nakagawa M et al 1998 Role of IQGAP1, a target of the small GTPases Cdc42 and Rac1, in regulation of E-cadherin-mediated cell-cell adhesion. Science 281:832–835

Marie H, Pratt SJ, Betson M et al 2003 The LIM protein Ajuba is recruited to cadherin-dependent cell junctions through an association with alpha-catenin. J Biol Chem 278:1220–1228

McNeill H, Ozawa M, Kemler R, Nelson WJ 1990 Novel function of the cell adhesion molecule uvomorulin as an inducer of cell surface polarity. Cell 62:309–316

Mohler PJ, Gramolini AO, Bennett V 2002 Ankyrins. J Cell Sci 115:1565–1566

Nieset JE, Redfield AR, Jin F et al 1997 Characterization of the interactions of alpha-catenin with alpha-actinin and beta-catenin/plakoglobin. J Cell Sci 110 (Pt 8):1013–1022

Ozawa M 1998 Identification of the region of alpha-catenin that plays an essential role in cadherin-mediated cell adhesion. J Biol Chem 273:29 524–29 529

Ozawa M, Ringwald M, Kemler R 1990 Uvomorulin-catenin complex formation is regulated by a specific domain in the cytoplasmic region of the cell adhesion molecule. Proc Natl Acad Sci USA 87:4246–4250

Pokutta S, Weis WI 2000 Structure of the dimerization and beta-catenin-binding region of alpha-catenin. Mol Cell 5:533–543

Pokutta S, Drees F, Takai Y, Nelson WJ, Weis WI 2002 Biochemical and structural definition of the l-afadin- and actin-binding sites of alpha-catenin. J Biol Chem 277:18 868–18 874

Pradhan D, Lombardo CR, Roe S, Rimm DL, Morrow JS 2001 α-Catenin binds directly to spectrin and facilitates spectrin-membrane assembly in vivo. J Biol Chem 276:4175–4181

Ridley AJ, Schwartz MA, Burridge K et al 2003 Cell migration: integrating signals from front to back. Science 302:1704–1709

Rimm DL, Koslov ER, Kebriaei P, Cianci CD, Morrow JS 1995 Alpha 1(E)-catenin is an actin-binding and -bundling protein mediating the attachment of F-actin to the membrane adhesion complex. Proc Natl Acad Sci USA 92:8813–8817

Sako Y, Nagafuchi A, Tsukita S, Takeichi M, Kusumi A 1998 Cytoplasmic regulation of the movement of E-cadherin on the free cell surface as studied by optical tweezers and single particle tracking: corralling and tethering by the membrane skeleton. J Cell Biol 140:1227–1240

Takeichi M 1995 Morphogenetic roles of classic cadherins. Curr Opin Cell Biol 7:619–627

Tao YS, Edwards RA, Tubb B et al 1996 beta-Catenin associates with the actin-bundling protein fascin in a noncadherin complex. J Cell Biol 134:1271–1281

Vasioukhin V, Bauer C, Yin M, Fuchs E 2000 Directed actin polymerization is the driving force for epithelial cell-cell adhesion. Cell 100:209–219

Watabe-Uchida M, Uchida N, Imamura Y et al 1998 alpha-Catenin-vinculin interaction functions to organize the apical junctional complex in epithelial cells. J Cell Biol 142:847–257

Weiss EE, Kroemker M, Rudiger AH, Jockusch BM, Rudiger M 1998 Vinculin is part of the cadherin-catenin junctional complex: complex formation between alpha-catenin and vinculin. J Cell Biol 141:755–764

Yonemura S, Itoh M, Nagafuchi A, Tsukita S 1995 Cell-to-cell adherens junction formation and actin filament organization: similarities and differences between non-polarized fibroblasts and polarized epithelial cells. J Cell Sci 108:127–142

DISCUSSION

D Lane: The implication of your model is an exclusivity about α-catenin's interaction. It is either interacting with β-catenin or it is interacting with actin. Do you have any direct chemical proof of competition, or any sense structurally of why they are competing with each other?

Nelson: The β-catenin binding site on α-catenin is at the N-terminus while the actin binding domain is at the C-terminus: they are at opposite ends of the molecule. The type of experiment that would be interesting is as follows. If we preassemble the cadherin/β-catenin/α-catenin complex and then start to add actin, how much of α-catenin dissociates from the cadherin complex and binds actin? Under those conditions we do see α-catenin coming off the complex and associate with actin. We don't yet understand what is going on with the competition between those proteins, given that the binding sites are at opposite ends of the molecule. I suggest that one of the things that may be happening is that differences in concentration of the α-catenin solution may be regulating the equilibrium of the protein binding to the cadherin complex and coming off again. The formation of cell–cell contacts through cadherin leads to α-catenin accumulating at relatively higher concentrations on the plasma membrane than somewhere else on the cell surface or in the cytoplasm. Hence this is more likely to effect actin dynamics locally than somewhere else in the cell.

D Lane: It is a bit like a local diffusion-type model.

Nelson: Yes. The FLIP (fluorescence loss in photobleaching) experiment indicates that this group of proteins together with vinculin are incredibly dynamic in cells even 36 h after cell–cell contacts have formed and there is strong adhesions between the cells.

Sheetz: In terms of the FRAP (fluorescence recovery after photobleaching), looking from above, everything is overlapping and it will look like a random diffusion process, when there may be active movements into this region. Have you looked in more detail to see whether some of these movements are active, which would give you an artificially high diffusion coefficient?

Nelson: What do you mean by active?

Sheetz: A directed migration.

Nelson: I'm not sure whether I would call this an artefact, but this is what we have indicated. In real-time analysis of the intensity of the fluorescence of cadherin and β-catenin across the membrane at a contact site, we see a decrease in fluorescence up to eight microns either side of the FRAP spot. This is accompanied by an increase in fluorescence in the middle part.

Sheetz: That could be diffusion.

Nelson: I think it is. This is exactly what is happening to these proteins in the plane of the lipid bilayer as they are moving around dynamically.

Sheetz: The other possibility is that in some way they are tied to an active movement. Myosin could be dragging those components through the membrane in a directed fashion.

D Lane: The model for active movement would have to be that he selects a spot and now you have active movement into that spot. I can't see that model. With diffusion it seems a lot easier to explain coming in from both sides.

Nelson: Of course, I think this detail is interesting, but the point is that cadherin and β-catenin are still highly mobile, but with dynamics different from that of actin. This is the type of argument we have been trying to build.

Drubin: What is the biological relevance of the α-catenin phosphorylation? Does it occur in response to cell contacts?

Nelson: If we isolate a cadherin/catenin complex off the membrane of a cell such as this, having labelled with ^{32}P, then both β- and α-catenin are phosphorylated (our unpublished results). This is a serine/threonine phosphorylation. There are about 25 phosphoserine sites in α-catenin, spread throughout the protein, but we do not have any evidence that it is required for complex formation with cadherin and β-catenin.

Firtel: Does the concentration of α-catenin required *in vitro* to inhibit actin polymerization match up with the concentration of free non-phosphorylated α-catenin in cells? Is this what you would expect? If you knock off α-catenin from β-catenin with a fragment from β-catenin so the end is free, do you change the actin dynamics in the cell? Do you have a sense that this looks fine, but if you then go and ask biochemistry on the other side, does it coincide?

Nelson: Good question. The amount of α-catenin we are adding in our pyrene actin assays is similar to what other people have used for actin-binding proteins. I can't tell you the relative amount of α-catenin in the cell. We are now looking at this, but it is a large pool relative to what is on the plasma membrane. There is a caveat: I do think that the concentration of cadherins at the membrane through clustering will locally change the concentration of α-catenin in that spot. I don't know how to calculate this.

Firtel: Is phosphorylated α-catenin associated with β-catenin?

Nelson: Yes.

Firtel: Then the question is, is there free phosporylated α-catenin, or is the phosphorylated form all bound?

Nelson: We do not have much information on this.

Hong: Has anyone cross-linked the cadherin–catenin complex? In immunoprecipitation experiments does anything come down?

Nelson: Yes, we published this in 1994 (Hinck et al 1994). There the stoichiometry is one E-cadherin to one β-catenin to about 0.8 α-catenin.

Weijin: How do you interpret these data?

Nelson: I interpret these data to show that this complex exists. In our hands it doesn't exist with actin.

Borisy: Don't your experiments predict a labile complex with strong binding between cadherin and β-catenin, and weaker dynamic binding with α-catenin? If you ran this on a gradient you would find a peak with cadherin and β-catenin, and then a smaller peak including the α. Then if you added actin you would displace the α away, leaving β and cadherin.

Nelson: The problem with that experiment is that you are taking the E-cadherin/β-catenin/α-catenin away from the pool of α-catenin, thus diluting it. Under those conditions you are probably right about what we would see, in the sense that you have now changed the equilibrium completely.

Borisy: More importantly, you have challenged our notions about what is going on at the cadherin/α-catenin/β-catenin complex, and you have challenged the association with actin. Now you are trying to reconstitute the complex and figure out how it really works. It is worth mentioning additional molecular players. You have mentioned the vinculin molecule, and that α-catenin does not bind to full-length vinculin, but does bind to the head domain. There is a cryptic association.

Nelson: Vinculin is a problem, because you have to activate it, if this is what you are getting at. We have done this in the presence of either phosphatidylinositol-4,5-bisphosphate (PIP2) or actin, which are the two things that I know activate vinculin. We get the same result.

Borisy: When α-catenin is bound to the head domain of vinculin, does this change its association properties with actin?

Nelson: Without cadherin/β-catenin? We haven't done that experiment. But if you take cadherin, β-catenin, α-catenin and the head domain of vinculin, this still does not bind to actin.

Braga: What do you think is going on in the very fast actin recruitment when you cluster cadherin receptors? Do you think there is another transmembrane protein that is co-clustered which recruits actin directly?

Nelson: I am actually very confused by this statement. The actin dynamics to me look like the actin falls apart under the contact. We see cortical actin as everyone sees for migratory cells, and as they come into contact that cortical organization of actin breaks down. What happens is that the ends of the actin bundles appear to go out towards the edge of a contact. We end up with actin like a horseshoe in each cell. Most of the actin we see by fluorescence seems to be associated with membrane dynamics. Lamellipodia and other structures which are sweeping over the cell, which you showed, are induced by activation of small GTPases.

Braga: One point is that the incorporated actin at newly formed cell–cell contacts is a very small fraction of the polymerized actin. I suspect when you have so much going on at the lamellae, you might not be able to detect this.

Nelson: I can't exclude what you are saying, but they look radically different from how they looked as the cells came into contact.

Kaibuchi: We previously calculated the amount of β-catenin and E-cadherin in MDCK cells. If they are regarded as soluble proteins, their concentrations might be around 1 μM. What percentage of α-catenin is free from β-catenin in your system?

Nelson: In whole cells the ratio we see is 1:1:0.8. The rest of it is free. I don't know precisely what this figure is at the moment.

Kaibuchi: In our hands, a lot of α-catenin appears to stay in the soluble fraction as the free form. The concentration may be about 1 μM in the cytoplasm. Is this concentration high enough for the inhibition of actin?

Nelson: Yes, there is a large pool. As I pointed out, if this model has some degree of sanity, at cell–cell contacts I would argue that the concentration of α-catenin would be higher than somewhere else in the cytoplasm. Even if we come up with a number for this, I am not sure how important it is relative to thinking about this local concentration.

Titus: Denise Montell's lab has shown that myosin VI binds to β-catenin, which in turn links to E-cadherin. Myosin VI apparently plays a role in adhesion and cell migration. Also there are links between a protein called vezatin and myosin VII. I don't know whether you have looked at any of these myosins in your cells, but they may be providing a link to the actin cytoskeleton that α-catenin may have to come in and undo.

Nelson: We haven't formally looked at myosins. We have taken a poor man's approach to this: in our reconstitution on membranes we added cytosol. This is because we were worried that we might be missing factor 'X'. The thought was that the cytosol would have factor 'X' and everything else to go with it. We still do not see actin polymerization. In addition we have tried to reconcile the biochemistry with what we have seen by live cell imaging using FRAP and FLIP and so forth. The numbers seem to fit fairly well with the models we are building from the biochemistry.

Chang: Seeing that a protein *in vitro* will inhibit actin assembly, doesn't mean it will work that way in the cell. Profilin is a good example of that. How do you link all your work up to formins? Bud6 and other formin binding proteins may be stimulating formin activity in actin assembly.

Nelson: There was a paper in *Nature Cell Biology* from the Fuchs lab (Kobielak et al 2004). In the two-hybrid system they found that formins bind α-catenin, and showed that formins were localized to points of cell–cell contact in keratinocytes. The key experiment was that on an α-catenin knockout background, they made a β-catenin–formin chimera. They said that this could rescue the adhesion. This is a forced interaction so I don't know whether it really happens. We can't get hold of a full-length formin to try to reconstitute this in our system. I would like to do this. Fred Chang, your point is well taken. That is why we tried to go into the whole cell. To design the experiment where we would test this model turns out to be difficult. We want to modify levels of the protein and look not just at actin polymerization and dynamics, but look at it immediately under the plasma membrane. These cells are fairly tall and polarized for 36 h, but there are significant optical constraints that stem from this.

Sheetz: On these blown-off membranes have you looked at FRAP?

Nelson: No. It's a good suggestion.

Sheetz: Have you looked at this lateral motility when you put them onto a basement membrane? Does it show the same motility as the MDCKs on glass?

Nelson: We have done this on filters where they accumulate a laminin/type IV collagen basement membrane. They look very similar.

Gundersen: Another approach to this problem is a structural one. I'm trying to reconcile your localized, but not directly attached actin scenario with electron micrographs (EMs) of adhesion contacts with actin filaments apparently inserting right into the adhesion contact. Can you comment on this?

Nelson: You would have to show me the paper you are talking about.

Borisy: Does such a reference exist?

Nelson: As Vania Braga says, there are people showing this lateral association of filaments. The perpendicular ones I am aware of are mostly from keratinocytes grown at low density—the Fuchs model of how a junction assembles.

Borisy: Have you done EM?

Nelson: No.

Borisy: It is clear that this is a missing part of the picture: high-resolution information about how the actin is organized at the junction. Light microscopy is not adequate for this.

D Lane: And it has to be done in your system, where you are not dealing with that movement at the same time. This is a potential cause of error.

Titus: In cross sections of epithelia at the sites of junctions, don't you see actin coming out perpendicularly?

Nelson: As Vania Braga indicated, she sees buttons of actin staining there. People interpret those as spikes coming out. We have looked pretty hard at this, so I would be interested if there was something out there.

B Lane: I have looked at quite a lot of electron micrographs of epithelial cells in tissues and in culture, but I have never seen any clear images of actin filaments running into an adhesion junction that come in perpendicular to the membrane.

Braga: One other comment is that when we look at mature junctions there are so many receptors there and they all bind to actin. We can see the bundles, but we don't know whether they are associated with adherens junctions or not. We need to reexamine this at high resolution.

Nelson: Gary Borisy, the EM you developed was key in understanding what was happening in lamellae.

Borisy: I agree!

Nelson: It was important. One of the things that will also be important will be to try to do EM at different times during cell–cell contact formation.

Humphries: I wonder whether assays that you are using to detect the complex are ever going to pick up certain K_ds. Are you using something like a Biacore or something more sophisticated than that? If you can measure K_{off} parameters for some of these interactions, maybe you can model the avidity effect.

Nelson: That is a great question. Bill Weiss with whom we collaborate is a structural biologist. Bill has measured the K_d using Biacore between E-cadherin and β-catenin which is 9 nM. There are significant problems doing the Biacore with β-catenin and α-catenin. They are now using calorimetry to try to work out the interaction. My guess is that it will be very weak. I think it is a good point about the equilibrium between these proteins. All our solution biochemistry is done at fairly high protein concentrations which will tend to force the interactions if they exist.

Humphries: You are right that the link hasn't been formally proven. It is very similar for integrins. I think it is a kinetic problem.

Nelson: This comes back to the issues about the dynamic equilibrium between these proteins. I am not a developmental biologist, but I am trying to reconcile how you maintain the integrity of the tissue and yet have tremendously dynamic movement.

Kaibuchi: There is a famous chimera molecule of E-cadherin fused to α-catenin. Can you make any comments on this?

Nelson: It is a nice experiment. Nagafuchi and colleagues made a chimera between the cytoplasmic domain of E-cadherin and α-catenin in a stable complex (Imamura et al 1999). Under these conditions the cells formed strong adhesions in a spinner culture. There is no evidence that this complex bound actin. The strong adhesion could be due to efficient delivery of E-cadherin to the cell surface. We have not made this chimera; we made the α-catenin–β-catenin chimera which doesn't bind actin in solution. I have no explanation other than to say that it may not need an actin interaction to do this.

Kaibuchi: When we express this chimera in L cells, we can't see the big accumulation of actin filaments in the cortical region.

Nelson: This goes back to the Steinberg models from 40 years ago thinking about the levels of cadherin on the cell surface. He showed that if you modulate the levels of different cadherins on the cell surface, this modifies the strength of adhesion between cells (Duguay et al 2003). It is well known that the cadherins will form crystalline arrays by aggregation. If the cells are coming into contact they are activating small GTPases such as Rac which are inducing membrane dynamics and the cells sending lamellipodia over each other. All of this can drive cadherins to this point to form a strong contact.

Vallee: What if you redid all your experiments with baculovirus-expressed proteins, given that you have micromolar levels of the protein? Could that make a difference in the observed reactions? We use bacterially expressed proteins with trepidation because we never know what activity is lacking.

Nelson: We haven't done this in baculovirus. I should point out that the proteins that we purify are the ones that Bill Weiss has crystallized, so we know they fold correctly as bacterially expressed proteins. In baculovirus one can do phosphorylation and so on, and that is why we have gone back to try to phosphorylate specific proteins. We added cytosol to see whether we can reconstitute any other types of activity, and we assume that cytosol will also contain kinases and phosphatases. We can't.

Peter: Why is your C-terminal deletion stable?

Nelson: There are two possibilities. One is that the truncated cadherin is intercalated into higher-ordered structures formed by the endogenous, full-length cadheirn. Second, it is known from studies published by Alpha Yap a number of years ago (Le et al 1999) that there is a lot of endocytosis and exocytosis of cadherin at the cell surface. The lack of the cytoplasmic domain may have some effect on that process as well.

Peter: You would think from the FRAP that it comes in laterally, and not really through endocytosis.

Nelson: That is correct. My speculation is that it is aggregating to form strong interactions on the membrane, and that is why it is very mobile. It is a surprising result.

Drubin: What effect does getting rid of α-catenin or actin have on the integrity of cell–cell junctions?

Nelson: We don't know for α-catenin. For actin, we have added cytochalasin and this has little to no affect on the dynamics of cadherin–catenin complex at cell–cell contacts.

Drubin: Do the adhesions form but not as strongly?

Nelson: In our hands that is what we see. It depends on how new the contacts are. If you take these cells which have been together for 36 h, cytochalasin doesn't do that much. However, if you take cells that have just formed a contact, in our hands many of these cells are migrating and they retract from each other. There are a lot of gross effects on the cell that are not directly to do with the cadherin contacts.

Gu: You mentioned that the strength between the junctions is dependent on the level of cadherin on the cell surface. You also mentioned that cadherin is regulated by endocytosis. What are the signals that trigger the endocytosis of cadherin then?

Nelson: I don't think they are known. It is a fact that cadherin is endocytosed, and that hakai can ubiquitinate cadherin and target it for degradation in the proteasome, but it is not known how those two events are regulated. Some people might even say that it is a constitutive process.

Gu: If it is just constitutive, it would be hard to regulate the tightness.

Nelson: If you think of an epithelial cell, the fact that I can't pinch off my skin is because the integrity of the tissue is maintained mostly by desmosomes and keratins. Even an MDCK cell, as much as a keratinocyte, will assemble other types of junctions, one of which is desmosomes. My view would be that the cadherins come in and initiate cell adhesion, and other proteins would come in secondarily such as desmosomal cadherins and keratins, to stabilize and strengthen those contacts.

B Lane: It takes 15–30 min to set up EM-recognizable desmosomes. It's very fast.

Nelson: The classical cadherins might be there to regulate the dynamics of those interactions.

D Lane: In these cells you are looking at, at the 36 h timepoint, are desmosomes already present?

Nelson: Yes.

D Lane: So presumably there is no structural need for cadherin junctions. You discussed issues about how cells move over each other: are these desmosome issues rather than cadherin junction issues?

Nelson: They are an issue for all junctions.

Borisy: In your presentation, you showed that cells in developing embryos move rapidly relative to each other. Based on this observation, it shouldn't be much of a surprise that the photobleaching results showed rapid recovery. Before any of the photobleaching or incorporation experiments were done in the field of the cytoskeleton we only had immunofluorescent snapshots. The presumption, even from the very term 'cytoskeleton', was that all the components were stable and slow to turn over. The first photobleaching and incorporation experiments were shocking. Are shockwaves now occurring in other fields where they are only now doing these experiments? So far it is a truism that any structural component in the cell turns out to be dynamic. This is one of the basic strategies of cell organization.

Gundersen: In muscle the actin and myosin in the sarcomere turnover very slowly and are presumably very stable.

Nelson: Even a cell like a keratinocyte, which you would argue is pretty stable, can be mobilized. If I cut myself the keratinocytes around the cut site with respond almost instantaneously through interleukins and other factors. They change their keratin expression and become more plastic and migratory. The cells become very motile to move into the wound site.

Humphries: If you have an interaction with a K_d of 10^{-7} M, which is a moderate interaction, the half-life of the complex can be as low as a minute. We shouldn't be surprised that complexes are breaking all the time.

D Lane: When I saw those cells move they reminded me of magnetic marbles, which can do that sort of movement and they are still in contact with each other. It is a question of allowing rapid exchange of contacts but maintaining a minimum number of contacts.

Nelson: The cells move very smoothly.

Sheetz: The only way that cells are going to generate forces needed to move relative to one-another in the plane is by pulling laterally on the adjacent cell.

B Lane: Most of these cell–cell junctions are mediated by cadherins, and each individual cadherin interaction is not very high affinity. The junctions are probably a bit like Velcro, with individual 'hooks' easily adhering and detaching but larger patches being more difficult to dislodge.

References

Duguay D, Foty RA, Steinberg MS 2003 Cadherin-mediated cell adhesion and tissue segregation: qualitative and quantitative determinants. Dev Biol 253:309–323

Hinck L, Nathke IS, Papkoff J, Nelson WJ 1994 Dynamics of cadherin/catenin complex formation: novel protein interactions and pathways of complex formation. J Cell Biol 125:1327–1340

Imamura Y, Itoh M, Maeno Y, Tsukita S, Nagafuchi A 1999 Functional domains of alpha-catenin required for the strong state of cadherin-based cell adhesion. J Cell Biol 144:1311–1322

Kobielak A, Pasolli HA, Fuchs E 2004 Mammalian formin-1 participates in adherens junctions and polymerization of linear actin cables. Nat Cell Biol 6:21–30

Le TL, Yap AS, Stow JL 1999 Recycling of E-cadherin: a potential mechanism for regulating cadherin dynamics. J Cell Biol 146:219–232

Integrin–syndecan cooperation governs the assembly of signalling complexes during cell spreading

Martin J. Humphries, Zohreh Mostafavi-Pour, Mark R. Morgan, Nicholas O. Deakin, Anthea J. Messent and Mark D. Bass

Wellcome Trust Centre for Cell-Matrix Research, Faculty of Life Sciences, University of Manchester, Michael Smith Building, Oxford Road, Manchester M13 9PT, UK

Abstract. Cell adhesion to fibronectin (FN) triggers the formation and maturation of adhesion complexes by modulating the activity of the Rho family of GTPases. Cells plated onto a ligand of integrin $\alpha5\beta1$ spread but fail to form focal adhesions or fully organize actin into bundled stress fibres unless co-stimulated with a ligand of syndecan 4. Engagement of syndecan 4 in such pre-spread cells recapitulates the Rac1 and RhoA activation profiles observed during spreading on whole FN. Furthermore, since adhesion to a ligand of $\alpha5\beta1$ alone does not activate Rac1, engagement of syndecan 4 appears to be an absolute requirement. In related work, we have examined differences in the mechanism of focal adhesion formation mediated by the FN-binding integrins $\alpha4\beta1$ and $\alpha5\beta1$. Two signalling differences were found. First, while $\alpha5\beta1$ required syndecan 4 as a co-receptor, $\alpha4\beta1$ did not. Second, focal adhesion formation via $\alpha5\beta1$ required PKCα activation, but only basal PKCα activity was observed following adhesion via $\alpha4\beta1$. These findings demonstrate that different integrins can signal to induce focal adhesion formation by different mechanisms.

2005 Signalling networks in cell shape and motility. Wiley, Chichester (Novartis Foundation Symposium 269) p 178–192

The fundamental importance of cell–extracellular matrix (ECM) adhesion for multicellular life has been established by a combination of genetic and pharmacological analyses (Bokel & Brown 2002, Danen & Sonnenberg 2003). Studies in model organisms have demonstrated that interactions between ECM assemblies and their cell surface receptors (primarily integrins and syndecans) provide physical support for tissues and a platform for directed migration. In patients with inflammatory, neoplastic and infectious diseases, aberrant adhesion perturbs cellular trafficking and causes dysregulation of cellular differentiation. Adhesion receptor function is partly determined by an ability to tether the contractile cytoskeleton to the plasma membrane, but there is also evidence that integrins and syndecans modulate signalling events that are essential for cellular differentiation (Arroyo et al 1999, Gardner et al 1999, Reizes et al 2001, Kaksonen et al 2002). There is now evidence for alterations in the fluxes of almost all known signalling pathways subsequent to adhesion, suggesting that the function of integrins and syndecans is integrated with other receptor systems.

In fibroblasts spreading on fibronectin (FN) *in vitro*, adhesion signalling complexes are distributed focally rather than diffusely, and appear as flecks, patches and stripes. These contact points are found all over the ventral surface, and are usually associated with the contractile polymers of the cytoskeleton. Morphological and functional analyses have defined three major forms of adhesion contact: focal complexes (FC), focal adhesions (FA) and fibrillar adhesions (FB; reviewed in Geiger et al 2001). These contacts are formed and disrupted in a sequential manner as cells move. They therefore reflect different stages of interaction of cells with the ECM, and presumably contain different collections of signalling molecules. FA are linked to actomyosin-containing stress fibres, and contain at least 50 different proteins (Zamir & Geiger 2001). The members of the Rho family of small GTPases play a central role in modulating the actin cytoskeleton: Cdc42 and Rac1 promote FC formation and membrane protrusions at the leading edge via filopodia and lamellipodia (Ridley et al 1992, Nobes & Hall 1995), while RhoA mediates FA formation and causes retraction of the trailing edge (Ridley & Hall 1992). Coordination between these pathways involves activation of p190RhoGAP which may permit membrane protrusion by suppressing RhoA activity (Arthur & Burridge 2001). Integrin occupancy is reported to control the activity of Rho family GTPases and regulate their translocation to plasma membrane microdomains (Del Pozo et al 2000, 2002, 2004).

Integrins undergo *cis* interactions with a number of different receptors, and thereby spatially regulate diverse signalling responses (reviewed in Giancotti & Tarone 2003). Direct extracellular associations with members of the TM4 family have been established, as has indirect co-clustering with a number of growth factor and cytokine receptors. It is notable that most ECM molecules possess both integrin- and syndecan-binding sites, and a clear synergistic relationship exists between these two families. Thus, $\alpha5\beta1$-dependent FA formation on FN requires engagement of, and signalling via, a syndecan co-receptor (Woods et al 1986, Bloom et al 1999). Of the four known members of the syndecan family of proteoglycans, only syndecan 4 has been found in FA (Woods & Couchman 1994). Treatment of cells with anti-syndecan 4 antibody triggers FA formation in cells adherent to an integrin-binding fragment of FN (Saoncella et al 1999) and fibroblasts from syndecan 4 knockout mice are unable to respond to FN (Ishiguro et al 2000). The importance of syndecan 4 for migration *in vivo* is exemplified by the wound healing and angiogenesis defects observed in null mice

(Echtermeyer et al 2001). The cytoplasmic domain of syndecan 4 binds to, and activates, PKCα, and the requirement for syndecan engagement for FA formation is bypassed by exogenous PKCα activation (Baciu & Goetinck 1995).

Particularly distinctive cellular responses have been observed on substrates recognized by the FN-binding integrin $\alpha 4\beta 1$. For example, $\alpha 4\beta 1$ engagement promotes enhanced cell migratory activity, while reducing spreading and FA formation. Studies employing integrin chimeras demonstrated that these functional properties were conferred by the $\alpha 4$ cytoplasmic domain (Chan et al 1992, Kassner et al 1995). This finding suggested that the cytoplasmic domain either modulates association of cytoskeletal and signalling molecules with its partner $\beta 1$ subunit differently to other $\beta 1$-associated α subunits, or interacts directly with cytoskeletal and signalling molecules. The latter hypothesis is supported by the fact that $\alpha 4$ can bind directly to paxillin, and that this association contributes to the reduction of spreading and promotion of migration (Liu et al 1999, Liu & Ginsburg 2000). Although the role of syndecans in $\alpha 4\beta 1$-mediated adhesion is unknown, it is intriguing that the binding sites for both molecules overlap within FN. The $\alpha 4\beta 1$-binding domain of FN is primarily located in the type III connecting segment (IIICS; Wayner et al 1989), which is adjacent to the major heparin-binding domain, termed HepII. Three sites each for integrin and heparin binding have been pinpointed within the HepII/IIICS region (Fig. 1). The overlapping locations of these sites suggest a close coordination between integrin and proteoglycan binding.

Results and discussion

Effect of syndecan 4 ligand binding on adhesion contact formation and maturation

Normal human fibroblasts were seeded onto the different FN fragments depicted in Fig. 1 and stained with anti-vinculin antibody as a marker of FA. As reported previously (Woods et al 1986), cells plated on a 50K fragment that encompasses type III repeats 6–10 and includes the $\alpha 5\beta 1$-binding site were unable to form FA unless stimulated with a soluble heparin/syndecan-binding fragment of FN (H/0; Fig. 2). Cells failed to spread on H/0 itself, indicating that the formation of FA is dependent upon cooperative signalling by the two types of receptor. Interestingly, when cells were pre-spread on 50K and the effect of H/0 was followed as a function of time, the formation and maturation of adhesion complexes was found to follow a distinct temporal sequence. Within 15 minutes of H/0 addition, FC were formed at the base of new lamellipodia/ruffles. By 30 minutes, these adhesions matured into FA that were located at the termini of actin stress fibres. Similar effects were observed following addition of an anti-syndecan polyclonal antibody (not shown). Taken together, these findings highlight similarities in the formation

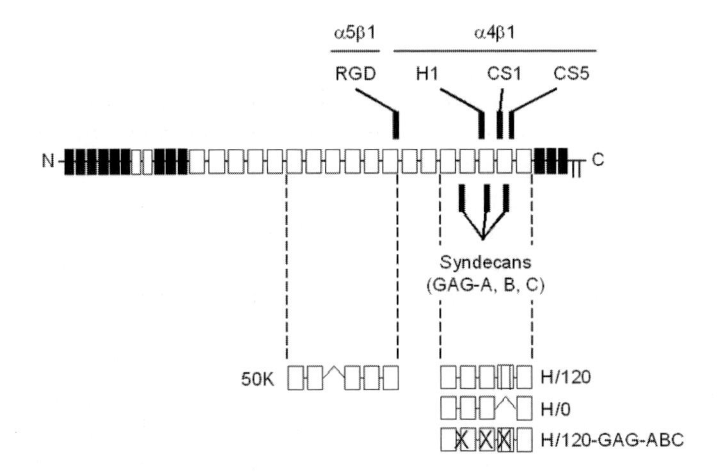

FIG. 1. Fibronectin (FN) comprises three types of polypeptide module termed types I, II (both white rectangles) and III (black rectangles). The central cell-binding domain, which binds $\alpha5\beta1$, contains an RGD tripeptide site in III-10. The alternatively spliced type III connecting segment (IIICS), which binds $\alpha4\beta1$, contains the active peptide sequences CS1 (in the N-terminal splice segment, IIICS-A) and CS5 (in IIICS-C). The main heparin-binding site fragment of FN (HepII) comprises III12-14. The main heparin-binding site in FN (GAG-A) is located in III-13, while additional heparin-binding sites are found in IIICS-B (GAG-B) and III-14 (GAG-C). HepII also contains an $\alpha4\beta1$-binding site, H1. The four FN fragments employed in these studies are 50K (III6-10), H/120 (III12-15, including a full-length 120-amino acid IIICS region), H/120-GAG-ABC (in which all heparin-binding sites are mutated), and H/0 (III12-15, without the IIICS).

and maturation of adhesion complexes observed either in response to clustering of syndecan 4 or during cell spreading on whole FN.

Regulation of Rac1 and RhoA activity by syndecan 4

Since the formation of FA and other adhesion contacts is regulated by members of the Rho family of small GTPases, we examined the extent of Rho family activation at different times following syndecan 4 engagement. An effector pull-down assay was employed in combination with GTPase Western blotting. 10 minutes after H/0 addition, when FC were being formed, Rac1 activity increased by ~50% (Fig. 3). After 30 minutes, Rac1 activity returned to basal levels. In this model system, Cdc42 activity did not change significantly. Levels of active RhoA were slightly, but highly reproducibly, suppressed at 10 minutes (Fig. 3). Following the initial suppression, activation of RhoA increased and peaked 60–90 minutes after stimulation. The identity of FC and FA was confirmed by blocking the RhoA effector Rho kinase using Y27632. This agent had no effect on formation of adhesion complexes at 10 minutes, but wiped out contacts seen at 60 minutes

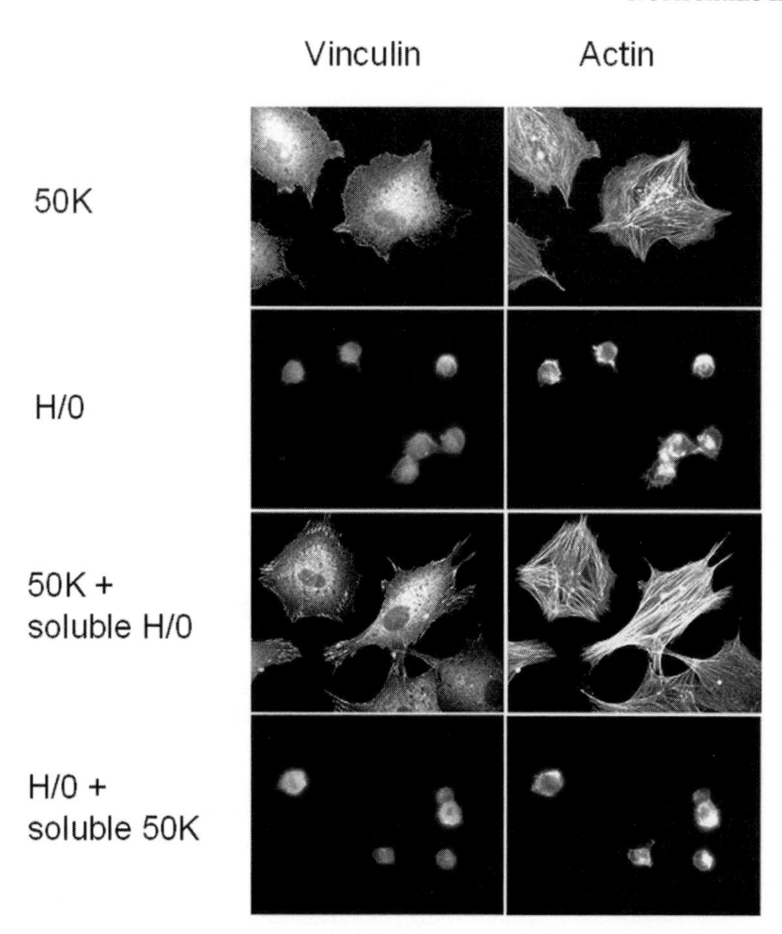

FIG. 2. Fibroblasts were seeded onto 50K or H/0, then allowed to spread for 2 hours in medium, supplemented with either 10 μg/ml soluble 50K or H/0 as indicated, before fixing and staining for vinculin and actin.

(not shown). Taken together, these observations demonstrate that ligand engagement by syndecan 4 regulates a cascade of GTPase activation with consequent effects on cell morphology. Furthermore, this cascade resembles the profile of GTPase activity observed during cell spreading on FN (del Pozo et al 2000).

Interdependence of Rac1 activation on integrin $\alpha 5\beta 1$ and syndecan 4

The similarity between the maturation of adhesion contacts observed during cell spreading on FN and following syndecan 4 engagement raises the question of

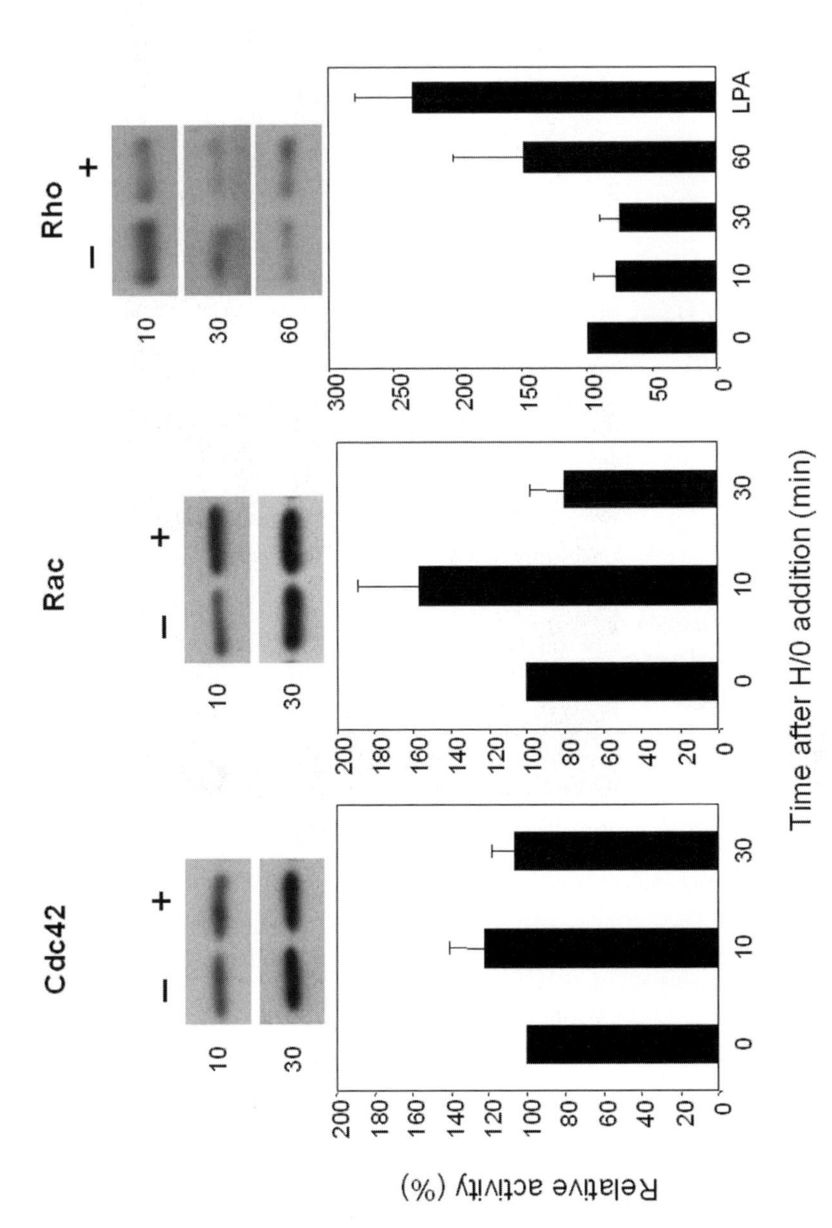

FIG. 3. Fibroblasts were allowed to spread on 50K for 2 hours and then H/0 was added for varying times. The time courses of activation of Cdc42, Rac1 and RhoA were determined by pull-down assays with GST-PAK-1 CRIB domain (Rac1 and Cdc42) or GST-rhotekin RBD domain (RhoA) followed by Western blotting. Equal loadings were confirmed by blotting crude lysates for total GTPase and vinculin.

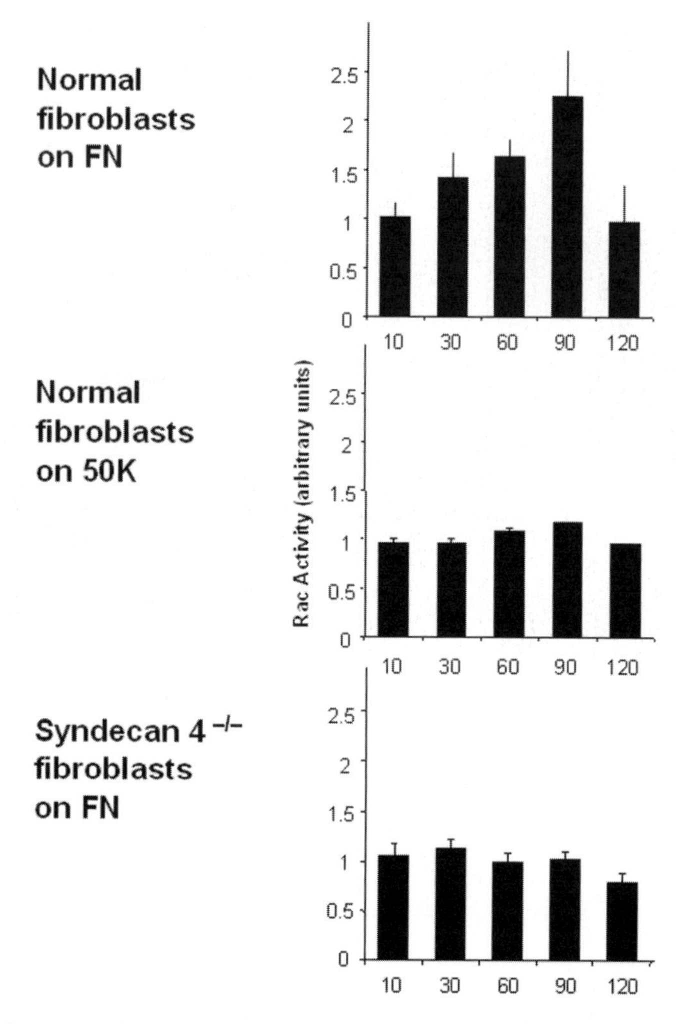

FIG. 4. Rac1 activation was determined during cell spreading of immortalized wild-type or syndecan 4 null mouse embryonic fibroblasts. Cells were plated onto either FN or 50K as indicated.

whether integrins are able to activate Rac1 in the absence of syndecan engagement. To test this possibility, we seeded normal fibroblasts on 50K and measured Rac1 activity as a function of time. As shown in Fig. 4, although the cells attached and spread, a wave of Rac1 activation was not observed. As a control, Rac1 activation did occur if the same cells were plated on whole FN. To obtain definitive proof that syndecan 4 contributes to Rac1 activation on FN, we generated syndecan 4 null

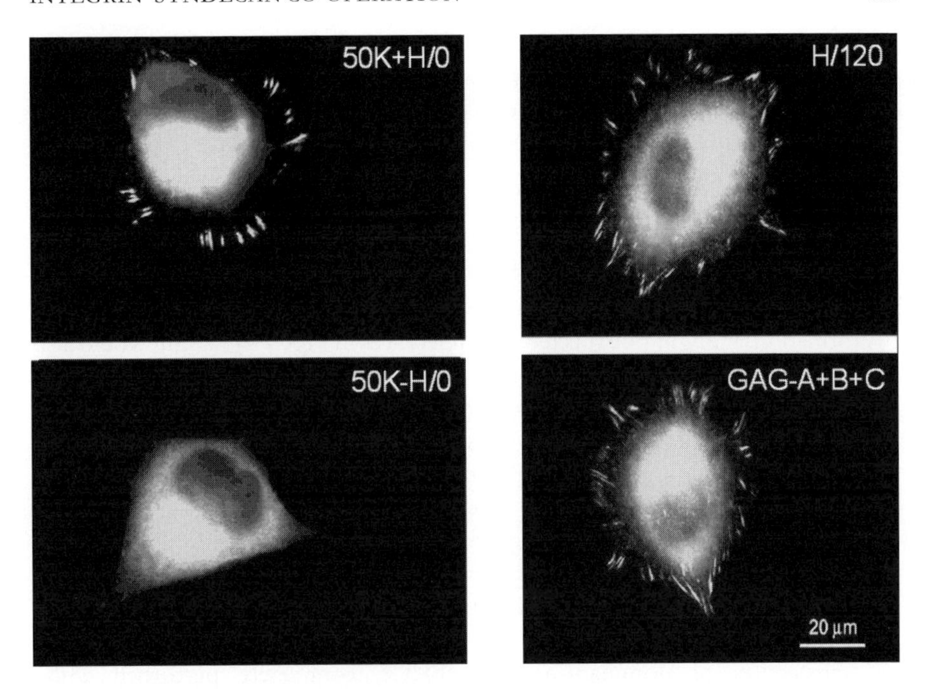

FIG. 5. A375-SM cells were incubated for 2 hours on 50K, 50K+soluble H/0, H/120 or H/120-GAG-ABC, fixed and stained for vinculin.

fibroblasts from knockout mice. When these cells were seeded onto whole FN, again there was no wave of Rac1 activation (Fig. 4). Thus, these findings demonstrate that syndecan clustering is obligatory for integrin-dependent Rac1 activation.

Requirement for syndecan engagement of different integrins

By employing fragments of FN that selectively engage different integrins, we next aimed to determine if the dependence of FA formation on syndecan engagement investigated above was a common feature of integrin signalling. Melanoma cells were selected for these studies because they express two FN-binding integrins, $\alpha4\beta1$ and $\alpha5\beta1$, as well as heparan and chondroitin sulfate-containing proteoglycans (Iida et al 1998). When A375-SM cells were seeded onto the H/120 fragment (Fig. 1), which binds both $\alpha4\beta1$ and syndecan 4, cells formed vinculin-containing FA within 60 minutes as expected (Fig. 5). Interestingly, when cells were seeded onto a mutant form of the same protein, H/120-GAG-ABC, which is unable to bind to syndecans, the cells adhered and formed FA that were indistinguishable from the wild-type protein (Fig. 5). The lack of dependence of $\alpha4\beta1$-dependent FA formation on syndecans was confirmed by seeding the cells onto CS1, which is a 25 amino acid peptide containing the main $\alpha4\beta1$-binding site in FN, but which lacks any detectable syndecan-binding activity (not shown; see Mostafavi-Pour et al 2003).

Contribution of PKCα activation to integrin-specific signalling

If $\alpha4\beta1$ and $\alpha5\beta1$ indeed differ in the signalling mechanisms employed to trigger FA formation, it might be expected that differences in PKCα signalling would also be observed when cells were employing either integrin for adhesion. Initially, since $\alpha4\beta1$ did not require a syndecan co-receptor to recruit vinculin, we hypothesized that PKCα activity levels would be constitutively high when cells were adhering via $\alpha4\beta1$. Surprisingly, the opposite was found. When cell lysates were Western-blotted with a site-specific antibody detecting phosphorylated PKCα, basal levels of active PKCα were observed that were similar to those induced by spreading on 50K (not shown; see Mostafavi-Pour et al 2003). By contrast, addition of H/0 to cells adhering via 50K induced an eightfold stimulation of PKCα. In complementary studies, addition of a pharmacological inhibitor of PKCα, BIM, or transfection of dominant-negative PKCα, completely perturbed $\alpha5\beta1$-dependent vinculin recruitment, but had no effect on recruitment by $\alpha4\beta1$. In conclusion, having compared the mechanisms of FA formation employed by melanoma cells when adhering to FN via $\alpha4\beta1$ and $\alpha5\beta1$, we conclude that:

- the two integrins differ in their requirement for syndecan engagement, with $\alpha5\beta1$, but not $\alpha4\beta1$, requiring ligand binding to a syndecan, and
- the two integrins trigger different signalling pathways that are distinguished by dependence on PKCα.

Future perspectives

The studies discussed above open up a number of interesting avenues for further exploration. First, the demonstration that syndecan 4 engagement can activate Rac1 implies that a signalling pathway exists linking the two molecules and the identity of the factors involved will be an important area to investigate. The syndecan cytoplasmic domain has been shown to bind a wide range of molecules, including ERM proteins, PKCα, PIP2, Src and PDZ domain-containing proteins (Bass & Humphries 2002). At the other end of the pathway, Rac1 activity is likely to be regulated by one or more guanine nucleotide exchange factors. Second, elucidating the links between this pathway and integrin signalling factors will hopefully provide an explanation of the syndecan dependence of $\alpha5\beta1$-mediated

FA formation. Third, as p190RhoGAP has been reported to mediate the initial suppression of RhoA activity during cell adhesion to FN, it will be informative to determine the effects of syndecan 4 engagement on both tyrosine phosphorylation and plasma membrane recruitment of p190RhoGAP, and then to identify the regulators of p190RhoGAP that depend on synergistic signalling between $\alpha5\beta1$ and syndecan 4. Fourth, the differences in proximal signalling events between $\alpha4\beta1$ and $\alpha5\beta1$ that determine the differential requirement for syndecan engagement and PKCα activity warrant examination. As $\alpha4\beta1$ has a number of unusual properties, it is conceivable that its 'gearing' to the cytoskeleton differs from other integrins. As the $\alpha4$ cytoplasmic domain binds paxillin, this is an obvious candidate to explain the differences in integrin signalling.

Acknowledgements

The work from this laboratory discussed within the article was supported by grants from the Wellcome Trust.

References

Arroyo AG, Yang JT, Rayburn H, Hynes RO 1999 $\alpha4$ integrins regulate the proliferation/differentiation balance of multilineage hematopoietic progenitors *in vivo*. Immunity 11:555–566

Arthur WT, Burridge K 2001 RhoA inactivation by p190RhoGAP regulates cell spreading and migration by promoting membrane protrusion and polarity. Mol Biol Cell 12:2711–2720

Baciu PC, Goetinck PF 1995 Protein kinase C regulates the recruitment of syndecan-4 into focal contacts. Mol. Biol Cell 6:1503–1513

Bass MD, Humphries MJ 2002 The cytoplasmic interactions of syndecan-4 orchestrate adhesion receptor and growth factor receptor signalling. Biochem J 368:1–15

Bloom L, Ingham KC, Hynes RO 1999 Fibronectin regulates assembly of actin filaments and focal contacts in cultured cells via the heparin-binding site in repeat III13. Mol Biol Cell 10:1521–1536

Bokel C, Brown NH 2002 Integrins in development: moving on, responding to, and sticking to the extracellular matrix. Dev Cell 3:311–321

Chan BMC, Kassner PD, Schiro JA, Byers HR, Kupper TS, Hemler ME 1992 Distinct cellular functions mediated by different VLA integrin α subunit cytoplasmic domains. Cell 68:1051–1060

Danen EH, Sonnenberg A 2003 Integrins in regulation of tissue development and function. J Pathol 200:471–480

Del Pozo MA, Price LS, Alderson NB, Ren XD, Schwartz MA 2000 Adhesion to the extracellular matrix regulates the coupling of the small GTPase Rac to its effector PAK. EMBO J 19:2008–2014

Del Pozo MA, Kiosses WB, Alderson NB, Meller N, Hahn KM, Schwartz MA 2002 Integrins regulate GTP-Rac localized effector interactions through dissociation of Rho-GDI. Nat Cell Biol 4:232–239

Del Pozo MA, Alderson NB, Kiosses WB, Chiang HH, Anderson RG, Schwartz MA 2004 Integrins regulate Rac targeting by internalization of membrane domains. Science 303: 839–842

Echtermeyer F, Streit M, Wilcox-Adelman S et al 2001 Delayed wound repair and impaired angiogenesis in mice lacking syndecan-4. J Clin Invest 107:R9–R14

Gardner H, Broberg A, Pozzi A, Laato M, Heino J 1999 Absence of integrin a1β1 in the mouse causes loss of feedback regulation of collagen synthesis in normal and wounded dermis. J Cell Sci 112:263–272

Geiger B, Bershadsky A, Pankov R, Yamada KM 2001 Transmembrane crosstalk between the extracellular matrix — cytoskeleton crosstalk. Nat Rev Mol Cell Biol 2:793–805

Giancotti FG, Tarone G 2003 Positional control of cell fate through joint integrin/receptor protein kinase signaling. Annu Rev Cell Dev Biol 19:173–206

Iida J, Meijne AM, Oegema TR et al 1998 A role of chondroitin sulfate glycosaminoglycan binding site in $\alpha4\beta1$ integrin-mediated melanoma cell adhesion. J Biol Chem 273:5955–5962

Ishiguro K, Kadomatsu K, Kojima T et al 2000 Syndecan-4 deficiency impairs focal adhesion formation only under restricted conditions. J Biol Chem 275:5249–5252

Kaksonen K, Pavlov M, Voikar V et al 2002 Syndecan-3-deficient mice exhibit enhanced LTP and impaired hippocampus-dependent memory. Mol Cell Neurosci 21:158–172

Kassner PD, Alon R, Springer TA, Hemler ME 1995 Specialized functional properties of the integrin α4 cytoplasmic domain. Mol Biol Cell 6:661–674

Liu S, Ginsberg MH 2000 Paxillin binding to a conserved sequence motif in the α4 integrin cytoplasmic domain. J Biol Chem 275:22736–22742

Liu S, Thomas SM, Woodside DG et al 1999 Binding of paxillin to α4 integrins modifies integrin-dependent biological responses. Nature 402:676–681

Mostafavi-Pour Z, Askari JA, Parkinson SJ, Parker PJ, Ng TTC, Humphries MJ 2003 Integrin-specific signaling pathways controlling focal adhesion formation and cell migration. J Cell Biol 161:155–167

Nobes CD, Hall A 1995 Rho, rac, and cdc42 GTPases regulate the assembly of multimolecular focal complexes associated with actin stress fibers, lamellipodia, and filopodia. Cell 81:53–62

Reizes O, Lincecum J, Wang Z et al 2001 Transgenic expression of syndecan-1 uncovers a physiological control of feeding behavior by syndecan-3. Cell 106:105–116

Ridley AJ, Hall A 1992 The small GTP-binding protein rho regulates the assembly of focal adhesions and actin stress fibers in response to growth factors. Cell 70:389–399

Ridley AJ, Paterson HF, Johnston CL, Diekmann D, Hall A 1992 The small GTP-binding protein rac regulates growth factor-induced membrane ruffling. Cell 70:401–410

Saoncella S, Echtermeyer F, Denhez F et al 1999 Syndecan-4 signals cooperatively with integrins in a Rho-dependent manner in the assembly of focal adhesions and actin stress fibers. Proc Natl Acad Sci USA 96:2805–2810

Wayner EA, Garcia-Pardo A, Humphries MJ, McDonald JA, Carter WG 1989 Identification and characterization of the T lymphocyte adhesion receptor for an alternative cell attachment domain (CS-1) in plasma fibronectin. J Cell Biol 109:1321–1330

Woods A, Couchman JR 1994 Syndecan 4 heparan sulfate proteoglycan is a selectively enriched and widespread focal adhesion component. Mol Biol Cell 5:183–192

Woods A, Couchman JR, Johansson S, Höök M 1986 Adhesion and cytoskeletal organization of fibroblasts in response to fibronectin fragments. EMBO J 5:665–670

Zamir E, Geiger B 2001 Molecular complexity and dynamics of cell-matrix adhesions. J Cell Sci 114:3583–3590

DISCUSSION

Firtel: If we look at the intracellular receptors or tyrosine kinases that may be involved in this, can you distinguish between any integrin response and syndecan response, and then see whether these are synergistic by trying to attack the intracellular pathway? There might be two different responses being activated, or they could be going through the same molecule, but you are getting a higher level of activation in order to mediate the response. It is difficult to make sense of the syndecans. It is not like adding soluble ligand where you can titrate out the response by adding increasing concentrations of ligand. Do you have any sense of what might be activated intracellularly to lead to the activation of the Rho? Presumably you are activating a tyrosine kinase. Have you been able to look at this and compare what the integrins and syndecans activate, to see whether they are different molecules and see how they might integrate?

Humphries: This hasn't been done for these receptors. But for other receptors Ken Yamada (Miyamoto et al 1996) has coated beads with different ligands or antibodies directed against different integrins. He looked at the recruitment of molecules to those beads. If you poison the cell with tyrosine kinase inhibitors then you drastically reduce the number of molecules that an integrin is able to recruit. He was able to identify a hierarchy of recruitment of molecules, depending on whether you had ligand there, or allowed or didn't allow tyrosine kinase activation. We know that focal adhesions contain many growth factor receptors, but we don't know whether there is a collaboration taking place that we are not probing in this system. I wouldn't be surprised if there was.

Sheetz: Where are the syndecans during these processes?

Humphries: That is a good question. There aren't many good antibodies directed against syndecans. We have only found one that stains cells well.

Sheetz: Can you put the GFP (green fluorescent protein) form there?

Humphries: We are trying this. The problem is that there is a PDZ domain at the end, so we can't put a GFP at the end. Guido David in Leuven has succeeded in inserting a GFP just inside the membrane. This works. We are also putting it at the N-terminus. The one antibody that does seem to work, which we have a small amount of, shows different things in different cells. In fibroblasts we can see syndecan 4 in focal adhesions that colocalize with integrins. The interesting story came with melanoma cells. We saw syndecan 4 in ruffling membrane. At the same time, the integrins were in focal adhesions, so there was a clear spatial separation of the two receptors. Is this something unique to melanoma cells or does it tell us something about the role of the syndecan? Since Rac activation triggers ruffling, is there a functional role for syndecans in the lamellipodium? The syndecan could have a particular function in rapid initial contact with a matrix because it is expressing highly mobile carbohydrates on its core protein.

Lim: Syndecans are proteoglycans. They can form an obstacle to migration. In the development of the nervous system, syndecans/proteoglycans serve as barriers for guidance of neurons, preventing cross-over from inappropriate sources. We have found that if we grow neuroblastoma cells on laminin, when they encounter the syndecan or proteoglycan barrier they don't cross. In terms of therapy for injured nerves, this response must be overcome; we can encourage neurites, the precursors of axons, to sprout but they won't cross into so-called glial scars. These scars are made of proteoglycans (such as syndecan) and tenascin, secreted by glia at sites of nerve injury. What are your views on overcoming the negative response to tenascin or proteoglycan barriers?

Humphries: I think there are two different anti-adhesive phenomena. If you just have a proteoglycan- or sugar-rich region, it may be that the cells don't stick if they only have a syndecan ligand. If there is a region that is devoid of integrin binding activity, this could happen. Tenascin works in a different way, by masking the integrin binding sites on fibronectin. The syndecan knockout mouse phenotype is interesting: the mice are viable but they have a wound healing defect. If you wound the skin of the mouse, the wound closes more slowly. This fits with a role in migration. It could be used for guidance, but I think they serve several functions. Syndecans are primarily characterized as being involved in growth factor presentation to receptors.

Gundersen: It has been important for you to separate the two binding sites to dissect the contributions of these two things. But do you think it is important in the intact fibronectin molecule that these binding sites are within a single fibronectin molecule, and that this may lead to different responses when the integrins and syndecans bind to an individual fibronectin molecule?

Humphries: I don't know whether anyone has really been able to address that issue. It is also an issue for whether multiple integrins can bind to the same fibronectin protein.

Manser: You didn't show any live cells spreading. If you have integrin matrix alone versus the combination, there is obviously a difference in the ability of the cells to spread, but does it look as though Rac is not being activated? Is there something different about the spreading process?

Humphries: This is the sort of morphology we see early on. We haven't really looked carefully to see whether there is protrusion and membrane ruffling. It is hard to believe that cells get to this stage without doing that. From the pull-downs you can see that there is a basal level of Rac activity. This is gross Rac activity from solubilized cells. What is really happening at the membrane may be more pertinent. It may be that what we are seeing in these cells at the bottom is inside–out driven spreading, and the cells are failing to respond to the outside.

Manser: If you look at the composition of the adhesion complex with a variety of antibodies, do you see any significant differences? You get adhesion complexes in both cases, but is there something different about the composition in each case?

Humphries: Yes, it is dramatic. In the first case you always get a minority of cells that do make some adhesion complexes, but 90% of them don't. Interestingly, we can see integrin being clustered but there is nothing with these clusters of integrins: no vinculin, actinin or paxillin, fascin or talin and so on.

Manser: It is not just that the adhesions are small and may be missing some components; they are just not forming then.

Humphries: We haven't seen anything associated with an integrin in this situation. Protein kinase C (PKC) seems to be important. Syndecan ligation triggers PKCα, but we don't know then what the next stage is that allows all these things to be recruited.

Manser: If you plate on the 50K protein but you add phorbol ester, then everything is OK?

Humphries: Yes.

Manser: So you get back adhesions.

Humphries: Yes.

Hong: Is there any physical interaction between integrin and syndecan? Is this interaction influenced by binding with fibronectin?

Humphries: Not really. There is one study from Jim McCarthy's lab (Iida et al 1998) showing that a cell surface proteoglycan binds to an integrin, but I don't know of any other study where people have seen a direct interaction. I tend not to think of most of the membrane components in the focal adhesion as binding to each other. They are recruited together by what is outside and inside the cell. I don't think it is necessary for those two to bind together. There is certainly no evidence for it. It is possible that we need a certain spatial proximity of those two receptors to make this link. This comes down to whether the link requires proximity or happens at a distance. We have evidence for both: in the fibroblasts they are together and in the melanoma cells they are separate. Until we get better reagents for imaging we won't be able to tell.

Peter: You showed that Rho-GAP is getting to the membrane. You made a link to phosphorylation. Do you think that phosphorylation is needed to get it to the membrane? What is phosphorylating the GAP?

Humphries: Src.

Peter: That would be at the membrane, I guess.

Humphries: We don't have any data on this. If that is true, then localization of Rho-GAP doesn't require phosphorylation. It is phosphorylated when we see it.

Alberts: The rate-limiting step in activating a GTPase is going to be exchange or GDP-release. I can't believe that in these systems the best-described regulation of RhoA is simply the activation of the Rho-GAP. There must be more going on, but the field points to just this one factor. What else do you think might be regulating RhoA and Cdc42? What is known about the exchange factors that might be governing the GTPase activity here?

Humphries: I don't know much about RhoA myself. We are just starting in this area. I have thought more about Rac, for obvious regions. One report from the

David lab shows that the cytoplasmic domain of syndecan 4 can bind to Tiam1. There might be a role here for direct binding of an exchange factor to syndecan 4 which could be responsible for activating Rac.

Alberts: Another thing that might be worth considering is Rnds which are, more or less, the anti-Rho. When you turn on MAP kinase signalling, one of the gene products that is turned on is the Rnds. They are the anti-Rho and will disrupt the stress fibre network in focal adhesions. Have you considered looking at those?

Humphries: I am aware of them, but I haven't considered looking at them.

Borisy: You said that the phenotype of the syndecan knockout mouse was viable but defective in wound healing. You also said there are four members of the syndecan family. Was this knockout specific of syndecan 4?

Humphries: Yes.

Borisy: What can you say about the redundant function of the other syndecans? What is the phenotype of the other knockouts and can you delete several together?

Humphries: I can only give a partial answer to that. The other three syndecans are expressed tissue specifically, whereas syndecan 4 is ubiquitous. 1 is epithelial, 2 is fibroblastic and 3 is neuronal. There's a lack of redundancy of function because of this. Syndecan 4 is the only one found in focal adhesions and implicated in the regulation of the cytoskeleton.

References

Iida J, Meijne AM, Oegema TR Jr et al 1998 A role of chondroitin sulfate glycosaminoglycan binding site in $\alpha 4\beta 1$ integrin-mediated melanoma cell adhesion. J Biol Chem 273:5955–5962

Miyamoto S, Teramoto H, Gutkind JS, Yamada KM 1996 Integrins can collaborate with growth factors for phosphorylation of receptor tyrosine kinases and MAP kinase activation: roles of integrin aggregation and occupancy of receptors. J Cell Biol 135:1633–1642

Formation of multicellular epithelial structures

Keith Mostov, Paul Brakeman, Anirban Datta, Ama Gassama, Leonid Katz, Minji Kim, Pascale Leroy, Max Levin, Kathleen Liu, Fernando Martin, Lucy E. O'Brien, Marcel Verges, Tao Su, Kitty Tang, Naoki Tanimizu, Toshiyuki Yamaji and Wei Yu

Department of Anatomy, and Department of Biochemistry and Biophysics, University of California, San Francisco, CA 94143-2140, USA

Abstract. The kidney is primarily comprised of highly polarized epithelial cells. Much has been learned recently about the mechanisms of epithelial polarization. However, in most experimental systems the orientation of this polarity is determined by external cues, such as growth of epithelial cells on a filter support. When Madin-Darby canine kidney (MDCK) cells are grown instead in a three-dimensional (3D) collagen gel, the cells form hollow cysts lined by a monolayer of epithelial cells, with their apical surfaces all facing the central lumen. We have found that expression of a dominant-negative (DN) form of the small GTPase Rac1 causes an inversion of epithelial polarity, such that the apical surface of the cells instead faces the periphery of the cyst. This indicates that the establishment of polarity and the orientation of polarity can be experimentally separated by growing cells in a 3D collagen gel, where there is no filter support to provide an external cue for orientation. DN Rac1 causes a defect in the assembly of laminin into its normal basement membrane network, and addition of a high concentration of exogenous laminin rescues the inversion of polarity caused by DN Rac1.

2005 Signalling networks in cell shape and motility. Wiley, Chichester (Novartis Foundation Symposium 269) p 193–205

Epithelia are the most fundamental type of cellular organization found in metazoa (O'Brien et al 2002, Zegers et al 2003). The mammalian kidney is a prototypical example of an epithelial organ (Dressler 2002). Most of the kidney consists of epithelial cells, which form the specialized tubules of the nephron and collecting system. Simple epithelial cells form monolayer sheets of cells that line surfaces and cavities, which are often tubular in shape. Physiologically, in the kidney and elsewhere epithelial cells separate compartments. They form a barrier that restricts the movement of most molecules across the barrier, but allows the regulated transepithelial movement of certain molecules and ions, both passively and by active transport. Different cell types in the nephron exhibit a particularly complex array of specific transporters. These selective barrier and transport functions are at the heart of renal function and derangements are involved in many diseases.

To perform these functions epithelial cells are highly polarized. One manifestation of this polarity is that their plasma membrane is divided into discrete domains (Mostov et al 2000, 2003). The apical surface faces the lumen of the cavity or tubule and is specialized for exchange of ions and molecules with the lumenal contents (in the case of the kidney, with the filtrate and urine). The lateral surface is specialized for interaction with adjacent cells, while the basal surface is specialized for interaction with the underlying basement membrane and connective tissue. The lateral and basal surfaces are more or less continuous and often considered together as 'basolateral'. The basolateral surface is separated from the apical surface by tight junctions (TJs) (Matter & Balda 2003). TJs function both as a 'gate' to prevent paracellular movement of molecules and ions across the monolayer, and as a 'fence' to prevent diffusion of proteins and exofacial lipids in the plane of the membrane and thereby maintain the distinction between apical and basolateral surfaces.

Most of the work on epithelial polarization has used epithelial cells grown on permeable filter supports, available commercially in premounted units, such as Transwells and Millicells (Simons & Fuller 1985, Lipschutz et al 2001). When grown on such porous supports, epithelial cells can obtain nutrients from their basolateral surface, much as they obtain nutrients *in vivo* from the bloodstream. Epithelial cells grown on permeable filters are therefore much better differentiated and polarized than cells grown on conventional solid surfaces, such as plastic tissue culture dishes or coverslips. Much of this work has used stable epithelial cell lines, especially MDCK. These cells form a uniform, well-polarized monolayer and have been the mainstay of epithelial polarization research. MDCK cells resemble adult distal tubule and collecting system cells. Although they are not perfectly representative of the full spectrum of cell types found in the nephron, they are a very good model for analysing the basic principles of polarization. For example, we first observed that syntaxins 3 and 4, which are SNARE proteins involved in targeting membrane traffic to the correct plasma membrane domain, are specifically localized in MDCK cells to the apical and basolateral surfaces, respectively (Low et al 1996, 1998). This has since been extended to all of the ~17 cell types identifiable along the renal tubule in the rat as well as many non-renal epithelial cells, suggesting that it is a basic principle of epithelial polarization (Li et al 2002). In contrast, syntaxin 2 is variably found in the apical and/or basolateral surfaces in different renal cell types, indicative of cell-type specific differences. My lab has made a great deal of use of MDCK cells, especially in studying polarized membrane traffic and transcytosis.

Recently, we have moved to growing MDCK cells in 3D gels of extracellular matrix (ECM), such as collagen I. Growth of epithelial cells in 3D gels has been used by a few investigators, notably Bissell, for many years and is now becoming much more popular (Hagios et al 1998, Walpita & Hay 2002). This more closely resembles the *in vivo* situation, where cells interact with ECM, rather than with an artificial support. When plated in 3D ECM gels, MDCK cells spontaneously form spherical cysts, comprising a monolayer of cells surrounding a hollow lumen. The cells are polarized, with their apical surfaces all facing the lumen (O'Brien et al 2002, Zegers et al 2003). This is a model for formation of a very simple multicellular structure, essentially a reductionist view of a tissue. In cross-section the cyst resembles the cross section of a tubule, i.e. a monolayer of polarized epithelial cells surrounding a lumen.

We and others have studied the behaviour of MDCK cells in collagen gels. When single MDCK cells are plated in collagen gel, the cells initially have little obvious polarity. The cells begin to divide and form a small mass of two-to-four cells (O'Brien et al 2002, Zegers et al 2003). Some polarization is apparent, in that small lumina start to form between the cells and apical plasma membrane markers are found at these lumina. As division continues, the mass of cells enlarges. Often there are multiple lumina in a single cyst, which usually coalesce into a single lumen. Cells in the interior of the mass usually die by apoptosis (Lin et al 1999, Debnath et al 2002).

Montesano and Orci first reported that treatment of these cysts with hepatocyte growth factor (HGF) causes the cysts to form branching tubules over a period of several days (Montesano et al 1991a,b). This is reminiscent of branching tubulogenesis that occurs in the kidney and many other epithelial organs, though it is much simpler. Like any *in vitro* system, it is not a perfect mirror of *in vivo* processes. Nevertheless, it is an excellent model for studying the basic processes underlying tubule formation (O'Brien et al 2002, Zegers et al 2003). In particular, the mechanistic analysis of the orientation of polarity described below would be extremely difficult to perform *in vivo*. HGF binds to Met, a receptor tyrosine kinase, which in turn signals through at least five different signalling pathways (Birchmeier & Gherardi 1998). Much work by many groups has been done on following these signalling events. We have concentrated instead on the downstream events by which these cells rearrange themselves into tubules. We developed methods for doing double and triple label immunocytochemistry with confocal microscopy on whole mounts of ~1 mm thick collagen gels containing cysts and tubule. Although this is time consuming (antibodies have to diffuse in and out of the thick gels), it has greatly facilitated all of our work.

When cysts are treated with HGF in the first 6–12 h some of the cells send out extensions, large pseudopod-like protrusions of their basolateral surface (Pollack et al 1997, 1998). These cells still retain at least a small basolateral surface and apical-basolateral polarity. After ~24 h some cells divide and leave the monolayer to form chains of cells, two-to-three cells long. These chains have lost epithelial apical–basolateral polarity and instead resemble the polarization of migrating cells. Chain formation requires cell division; moreover cells change their mitotic spindle axis so that one of the daughter cells leaves the monolayer (Yu et al 2003). With increasing division, cells form cords of cells, two-to-three cells thick. New lumina begin to appear between the cells and the cells begin to repolarize, forming an epithelial monolayer. Eventually, the lumina coalesce and become contiguous with the central lumen of the cyst. Overall, the process of repolarization of epithelia during tubulogenesis appears to be quite similar to that occurring during initial cyst development.

We investigated the role of Rac1, a member of the Rho family of small GTPases, in cyst formation. Initially, we expected a role for Rac1 in tubulogenesis, specifically extension and chain formation, since Rac1 is frequently associated with cell motility. Surprisingly, we found that Rac1 was involved in controlling the orientation of epithelial polarity. We expressed dominant negative (DN) Rac1 (Rac1N17), using a tetracycline-repressible system. Some years ago we adapted the tetracycline-repressible system of Bujard and colleagues to MDCK cells and developed a MDCK clone, T23, which shows particularly tight regulation (Barth et al 1997). We had given these cells to Tzuu-Shuh Jou in James Nelson's lab at Stanford, who used them to express DN Rac1 and other Rho family proteins under tight control (Jou et al 1998, Jou & Nelson 1998). We obtained the T23 DN Rac1 cells back from them.

We found that cells expressing DN Rac1 formed cysts with inverted polarity (O'Brien et al 2001). Apical plasma membrane markers and TJ were around the periphery of the cysts. There were little or no lumina and basolateral surface markers were on all surfaces of the cells that contacted other cells. Normally, the entire structure of an epithelial cell is polarized, not just its plasma membrane. For instance, the Golgi apparatus is normally localized between the nucleus and the apical surface. We found that the inversion of polarity in DN Rac1 cysts extended to the internal organization of the cells. The Golgi apparatus was still between the nucleus and the new apical surface, which now was at the periphery of the cyst.

It is important to note that the cells were not simply depolarized or randomly polarized. Rather the orientation of polarity was precisely changed by 180°. This makes a fundamental point that the presence of cell polarization and the orientation of polarization can be separated experimentally. To our knowledge this question had not been considered explicitly in epithelial cells previously. To use a mathematical analogy, cell polarization can be likened to a mathematical vector, which has both a magnitude and a direction. Cells have both a degree of polarization and an orientation of that polarization.

Our results also raise the question of how polarization is controlled in multicellular structures, essentially a question of integrative biology. In a cyst or tubule, not only are all the cells polarized but also they all point in the same direction with the apical surface facing the lumen. This coordination among cells is at the heart of how cells are arranged into tissues and organs.

An important clue came from the behavior of DN Rac1 cells when grown on filter supports, which was first reported by Nelson and colleagues (Jou et al 1998, Jou & Nelson 1998). On filters, these cells do not have inverted polarity. There are some changes, such as alterations in TJ function and somewhat shorter height, but overall the cells look fairly normal. We suggest that when cultured on filters or any other artificial support, the support itself gives the cell an over-riding external cue as to how to orientate its polarity (O'Brien et al 2002, Zegers et al 2003). The surface in contact with the filter becomes the basal surface and the opposite free surface facing the overlying medium becomes the apical surface. In other words, cells on a support are in an artificially anisotropic environment. In contrast, when single cells are plated in a 3D collagen gel, they are in an isotropic environment and do not have any over-riding external cues to orientate their polarity. Cells must break symmetry to become polarized. As soon as a single cell has divided to form two cells, the cell faces two different environments, i.e. the ECM and the neighbouring cell, and this can provide the external cue to orientate polarity. Moreover, unlike culture on a support which are intrinsically provided with a free surface, cells in 3D gels must establish a free, apical surface.

An often-stated view is that cell–cell contact, mediated by cell–cell adhesion proteins such as E-cadherin, is the primary determinant of epithelial polarization. Nelson and colleagues, in particular, had proposed that E-cadherin contacts set up a 'targeting patch' where newly synthesized molecules were delivered to the cell surface (Drubin & Nelson 1996, Yeaman et al 1999). This model is true in the sense that cell–cell contact is needed for cells to polarize. However, the orientation of polarization cannot be determined simply by cell–cell contact; contact with a ECM or an artificial support is essential, as demonstrated by several seminal studies (Rodriguez-Boulan et al 1983, Vega-Salas et al 1987, Ojakian et al 1990, Wang & Ojakian 1990). The TJ is located at one end of the lateral surface where the cells contact their neighbours. The problem of orientation of epithelial polarity is, in some sense, a matter of the cell deciding at which end of the lateral surface to place the TJ and correspondingly where to put the apical surface. When grown on a filter both wild-type and DN Rac1 cells place the TJ at the end of the cell opposite from the filter. In contrast, in collagen gels, wild-type cells place the TJ and apical surface towards the center, while DN Rac1 cells place the TJ and apical surface at the periphery.

Although genetic analysis in *Drosophila* and *Caenorhabditis elegans* has been very powerful in identifying and analysing molecules involved in epithelial polarity, these *in vivo* systems are not particularly well suited for analysis of orientation of polarity (Knust & Bossinger 2002). Most of the work on *Drosophila* epithelial polarization has used the initial epithelium that covers the embryo after cellularization. This epithelium rests on the underlying organism and has a free surface defined by the outside of the organism. In this sense, it resembles epithelial cells cultured on a filter, where orientation of polarity is defined by the external cue of the filter and the free surface in contact with the overlying medium. Although many mutations affect the ability of the *Drosophila* epithelium to polarize, we know of none which simply invert the polarity of this covering epithelium. In this respect a powerful advantage of the simplified system of MDCK cells in 3D ECM gels is that it allows us to remove most external cues and thereby dissect out the orientation of polarity separately from the establishment of polarity.

These data suggested to us that cell interaction with ECM may be important in orientation of polarity. Indeed, work from several groups had previously indicated a link between ECM interactions and polarization, particularly polarization of apical proteins (Rodriguez-Boulan et al 1983, Vega-Salas et al 1987, Ojakian et al 1990, Wang & Ojakian 1990). The basal surface of epithelial cells normally sits on a basement membrane, a specialized ECM structure formed by the epithelial cells and surrounding stromal cells. Basement membranes have three major classes of constituents, including laminins, type IV collagen and proteoglycans. We reasoned that DN Rac1 might be acting on the formation of the basement membrane, which could be involved in orientation of polarization. Previous work had shown that laminin was important for epithelial polarization and kidney development, but did not specifically address its potential role in orientation of polarization (Klein et al 1988, Colognato & Yurchenco 2000).

Laminin is normally assembled into a network by laminin receptors on the cell surface (Colognato & Yurchenco 2000). These receptors apparently act simply by concentrating the laminin near the cell surface, raising the local concentration high enough to allow laminin self-assembly. Laminin assembly is a supra-molecular process and is best assessed by light microscopy. When wild-type cysts are stained for laminin, a fine network pattern is seen surrounding the cyst. In contrast, DN Rac1 cysts had laminin in a few large aggregates (O'Brien et al 2001). We found that DN Rac1 did not interfere with the synthesis, secretion or stability of laminin, as DN Rac1 cysts had normal amounts of laminin by Western blot and this was almost entirely secreted (i.e. accessible to trypsin added outside the cyst). The diminished staining intensity of laminin in DN Rac1 cysts is probably due to poor accessibility of laminin in large aggregates to antibodies during immunofluorescence staining.

To test whether laminin assembly was functionally important to the orientation of polarity, we next added exogenous laminin to the DN Rac1 cysts (O'Brien et al 2001). This laminin was commercially purified and of an embryonic type (laminin 1), but it is rather similar to the major laminin species expressed in kidney (laminin 5). At the high concentration added, the exogenous laminin could self assemble, by-passing the need for assembly mediated by cell-surface receptors. Addition of exogenous laminin strikingly rescued the orientation of cyst polarity, with the apical surface now in its normal position facing the lumen. Apical plasma membrane markers, TJ markers and Golgi were all rescued. Subsequently, Mina Bissell, Ole Petersen and colleagues reported that laminin plays a similar role in orientation of polarity in mammary epithelial cells (Gudjonsson et al 2002). In this case, the laminin is secreted by myoepithelial cells.

References

Barth AIM, Pollack AL, Altschuler Y, Mostov KE, Nelson WJ 1997 NH_2-terminal deletion of β-catenin results in stable colocalization of mutant β-catenin with adenomatous polyposis coli protein and altered MDCK cell adhesion. J Cell Biol 136:693–706

Birchmeier C, Gherardi E 1998 Developmental roles of HGF/SF and its receptor, the c-Met tyrosine kinase. Trends Cell Biol 8:404–410

Colognato H, Yurchenco PD 2000 Form and function: the laminin family of heterotrimers. Dev Dyn 218:213–234

Debnath J, Mills KR, Collins NL et al 2002 The role of apoptosis in creating and maintaining luminal space within normal and oncogene-expressing mammary acini. Cell 111:29–40

Dressler G 2002 Tubulogenesis in the developing mammalian kidney. Trends Cell Biol 12:390

Drubin DG, Nelson WJ 1996 Origins of cell polarity. Cell 84:335–344

Gudjonsson T, Ronnov-Jessen L, Villadsen R et al 2002 Normal and tumor-derived myoepithelial cells differ in their ability to interact with luminal breast epithelial cells for polarity and basement membrane deposition. J Cell Sci 115:39–50

Hagios C, Lochter A, Bissell MJ 1998 Tissue architecture: the ultimate regulator of epithelial function? Philos Trans R Soc Lond B Biol Sci 353:857–870

Jou TS, Nelson WJ 1998 Effects of regulated expression of mutant RhoA and Rac1 small GTPases on the development of epithelial (MDCK) cell polarity. J Cell Biol 142:85–100

Jou T-S, Schneeberger EE, Nelson WJ 1998 Structural and functional regulation of tight junctions by RhoA and Rac1 small GTPases. J Cell Biol 142:101–115

Klein G, Langegger M, Timpl R, Ekblom P 1988 Role of laminin A chain in the development of epithelial cell polarity. Cell 55:331–341

Knust E, Bossinger O 2002 Composition and formation of intercellular junctions in epithelial cells. Science 298:1955–1050

Li X, Low SH, Miura M, Weimbs T 2002 SNARE expression and localization in renal epithelial cells suggest mechanism for variability of trafficking phenotypes. Am J Physiol Renal Physiol 283:F1111–F1122

Lin HH, Yang TP, Jiang ST, Yang HY, Tang MJ 1999 Bcl-2 overexpression prevents apoptosis-induced Madin-Darby canine kidney simple epithelial cyst formation. Kidney Int 55:168–178

Lipschutz JH, O'Brien LE, Altschuler Y et al 2001 Analysis of membrane traffic in polarized epithelial cells. Curr Protocols Cell Biol, Vol 15.5

Low S-H, Chapin SJ, Weimbs T et al 1996 Differential localization of syntaxin isoforms in polarized MDCK cells. Mol Biol Cell 7:2007–2018

Low S-H, Chapin SJ, Wimmer C et al 1998 The SNARE machinery is involved in apical plasma membrane trafficking in MDCK cells. J Cell Biol 141:1503–1513

Matter K, Balda MS 2003 Signalling to and from tight junctions. Nat Rev Mol Cell Biol 4:225–237

Montesano R, Schaller G, Orci L 1991a Induction of epithelial tubular morphogenesis *in vitro* by fibroblast-derived soluble factors. Cell 66:697–711

Montesano R, Matsumoto K, Nakamura T, Orci L 1991b Identification of a fibroblast-derived epithelial morphogen as hepatocyte growth factor. Cell 67:901–908

Mostov KE, Verges M, Altschuler Y 2000 Membrane traffic in polarized epithelial cells. Curr Op Cell Biol 12:483–490

Mostov KE, Su T, ter Beest M 2003 Polarized epithelial membrane traffic: conservation and plasticity. Nat Cell Biol 5:287–293

O'Brien LE, Jou TS, Pollack AL et al 2001 Rac1 orientates epithelial apical polarity through effects on basolateral laminin assembly. Nat Cell Biol 3:831–838

O'Brien LE, Zegers MMP, Mostov KE 2002 Opinion: Building epithelial architecture: insights from three- dimensional culture models. Nat Rev Mol Cell Biol 3:531–537

Ojakian GK, Schwimmer R, Herz RE 1990 Polarized insertion of an intracellular glycoprotein pool into the apical membrane of MDCK cells. Am J Physiol 258:C390–C398

Pollack AL, Barth AI, Altschuler Y, Nelson WJ, Mostov KE 1997 Dynamics of beta-catenin interactions with APC protein regulate epithelial tubulogenesis. J Cell Biol 137:1651–1662

Pollack AL, Runyan RB, Mostov KE 1998 Morphogenetic mechanisms of epithelial tubulogenesis: MDCK cell polarity is transiently rearranged without loss of cell-cell contact during scatter factor/hepatocyte growth factor-induced tubulogenesis. Dev Biol 204:64–79

Rodriguez-Boulan E, Paskiet KT, Sabatini DD 1983 Assembly of enveloped viruses in Madin-Darby canine kidney cells: polarized budding from single attached cells and from clusters of cells in suspension. J Cell Biol 96:866–874

Simons K, Fuller SD 1985 Cell surface polarity in epithelia. Annu Rev Cell Biol 1:243–288

Vega-Salas DE, Salas PJ, Gundersen D, Rodriguez-Boulan E 1987 Formation of the apical pole of epithelial (Madin-Darby canine kidney) cells: polarity of an apical protein is independent of tight junctions while segregation of a basolateral marker requires cell-cell interactions. J Cell Biol 104:905–916

Walpita D, Hay E 2002 Studying actin-dependent processes in tissue culture. Nat Rev Mol Cell Biol 3:137–141

Wang AZ, Ojakian GK 1990 Steps in the morphogenesis of a polarized epithelium. II. Disassembly and assembly of plasma membrane domains during reversal of epithelial cell polarity in multicellular epithelial (MDCK) cysts. J Cell Sci 95:153–165

Yeaman C, Grindstaff KK, Nelson WJ 1999 New perspectives on mechanisms involved in generating epithelial cell polarity. Physiol Rev 79:73–98

Yu W, O'Brien LE, Wang F et al 2003 Hepatocyte growth factor switches orientation of polarity and mode of movement during morphogenesis of multicellular epithelial sructures. Mol Biol Cell 14:748–763

Zegers MMP, O'Brien LE, Yu W, Datta A, Mostov KE 2003 Epithelial polarity and tubulogenesis *in vitro*. Trends Cell Biol 13:169–176

DISCUSSION

Harden: We have been doing some work on Pak and Rac in development of the follicular epithelium surrounding the *Drosophila* egg chamber. The model you presented was the idea that the information from the basement membrane is being transmitted across the cell, possibly by Cdc42.

Mostov: In this case, but this is not necessarily universally true. What is universally true is that there has to be a cue.

Harden: If we make a Rac mutant clone in the follicle cells, the polarity is disrupted. The unusual feature of the ovariole is that there are inputs from the nurse cells at the apical surface, so there may be influences that you wouldn't have in your system. Interestingly, Pak mutant clones also lose polarity. The Pak kinase is specifically localized to the basal end of those cells, and can't be seen anywhere else in the cell. It collects at the basal end. There is an interesting possibility—as you know there are two other protein complexes which are required for apical–basal polarity: the Crumbs (Crb) complex and the more basolaterally localized Scrib/Lgl complex. There's a recent paper that indicates that Rac and Pix can bind Scribble (Audebert et al 2004). The model we are thinking of is that the basal surface could be impinging on the Scrib/Lgl complex. There are several proteins in this complex, including Scrib, Lgl and Discs large (Dlg). This complex appears to be able to contribute to the apical positioning of the Par6 atypical PKC complex that Keith Mostov was mentioning. This is another way to impinge on the apical localization of atypical PKC he was talking about.

Mostov: The models coming from Bilder and Perrimon are based on genetic interactions that are really quite complicated. The simplest view is that the Par3 complex is upstream, and downstream of it is the Crb complex apically and the Scribble complex basolaterally. The Scribble complex and Crb complex are more-or-less in competition pushing against each other. It is more complicated than this. My way of getting into this was as a membrane traffic person. There is some evidence, particularly from yeast, that this Lgl complex is involved in vesicular docking. It has been hard to pin this down in mammalian systems. It might be an over-simplification to imagine that there is the issue of attaining polarity one way, and these subsequent factors downstream help maintain that polarity and decide how big one surface will be versus the other. One point I would raise from all this is that we have made tremendous advances on the genetic dissection of these complexes. In almost all the mutations, though, we see many that will disrupt the epithelial polarity but people generally don't see an inversion of polarity. I would contend that the reason for that is that in these experimental systems there are still external constraints and cues. The epithelial cells still cover the interior of the organism and this provides the cue. People are never operating in an isotrophic environment. In contrast, the MDCK system we can get an isotrophic environment and so we can begin to see this inversion of polarity.

Borisy: You mentioned that these balls of cells in the isotropic collagen environment break symmetry and make a lumen normally. What is known about how they do this? They must be detecting something.

Mostov: Obviously, they are detecting collagen.

Borisy: I am not talking about the inversion of the symmetry. I am just talking about the normal process. How do they make a lumen?

Mostov: We don't understand this. There is contact through E-cadherin, although James Marrs and W. James Nelson showed that MDCK cells expressing a dominant-negative E-cadherin would still form perfectly polarized cysts (Troxell et al 2000). Cell–cell contact is probably needed, perhaps mediated by nectins and/or E-cadherin. Once a cell divides, even at a two-cell stage, we can see the earliest accumulation of apical markers in normal situations. There is the whole issue of how this apical surface is made. Presumably there is vesicular targeting of anti-adhesive and mucinous proteins. For example, the apical protein Gp135 probably has a lot of carbohydrate branching which may be somewhat anti-adhesive.

Borisy: So you have a paradoxical situation where adhesion is stimulating anti-adhesion. Is that what you are saying?

Mostov: Yes. It is almost teleological. The cells have a mechanism that is driving them to have three surfaces: one contacting matrix, one contacting other cells and a free surface. They have an intrinsic capability for creating a free surface. The phrase intrinsic capability means black box. We'd like to understand this.

Nelson: Joan Brugge's lab has used MCF7 cells which form acinar-type structures in collagen with a layer of cells around a fluid-filled space. They have shown that the cells in the centre undergo apoptosis and die.

Mostov: When we get large structures like this you need apoptosis. But this is not needed for lumens made at a two or four cell stage.

Nelson: Many years ago we showed that if these cells are denied a free surface, then the apical membrane vesicles tend to accumulate and aggregate. Rodriguez-Boulan's lab published a structure they called VAC, a vacuolar apical compartment, which is probably a very large exocytic structure of apical membranes which have fused with themselves but have nowhere to go in terms of a free surface (Vega-Salas et al 1987). We and they have shown that these structures can (probably stochastically) bump into the plasma membrane and fuse, and thereby generate a free surface. As Keith indicated, this would be a preformed apical membrane lacking all of these associated proteins that would be found on a basal or lateral surface.

Borisy: Let's be clear. Joan Brugge has shown that cells that enter the lumen of these acinar structures do undergo apoptosis, but apoptosis is not the explanation for producing the lumen.

Nelson: By default it would do.

Borisy: Are you suggesting a model in which the lumen is made by the death of cells?

Nelson: No, Joan Brugge is suggesting this model.

Mostov: Gail Martin and Electra Coucouvanis first showed this (Coucouvanis & Martin 1995). Brugge explored the mechanisms of apoptosis in mammary cells. In between, the person ignored altogether was Ming-Jer Tang, who discovered the apoptosis business in MDCK cysts (Lin et al 1999). Apoptosis is needed to create space in a large structure but it isn't needed to create the small lumen seen with two, four or six cells.

Manser: It is fascinating that you see MAP kinase activation or Raf activation with the hepatocyte growth factor (HGF), yet your cells are exposed to serum that presumably causes MAP kinase activation anyway.

Mostov: They have an endogenous level.

Manser: Is it suppressed by the fact that they have reached a certain stage, and then when you add HGF there is a stimulating signal?

Mostov: That is an interesting idea. I don't know. We haven't measured phospho-ERK kinase activity during the development of the cyst. What you are suggesting is that on a per cell basis this probably goes down. This sounds reasonable.

Manser: There is a parallel with the story of neuronal differentiation. You need a certain level of MAP kinase activation in order to get cell division, but if you push it high enough they differentiate.

Mostov: When we treat with HGF we see enormously prolonged ERP activation. If we add the U126 compound it will still go through the whole program, so we only need the MAP kinase cascade for the first 24 h.

Alberts: One thing the MAP kinase stimulation will do will be to lead to new gene expression. Have you considered what possible changes in gene expression might be driving this?

Mostov: I don't have any data. A former postdoc in the lab has been looking at some canine chips. He has published what happens when HGF is added on cells. This is what led us to MMP13, which is the gene on the chip that goes up the most. The limitation is that the canine chip is pretty small. There are disadvantages in looking at MDCK cells.

Luca: When you have these cysts with inverted polarity, if you add HGF, presumably this won't stimulate tubule formation but it will stimulate migration. Is that the case?

Mostov: They just sort of sit there.

Luca: Have you looked at the spindle orientation in those cells?

Mostov: No, the inverted cysts will not form tubules.

Borisy: Perhaps you could connect the issue of spindle orientation with some of the previous discussions we have had on microtubule capture. This is clearly related to the phenomenon of orientation of the spindle.

Mostov: I imagine that in monolayers like this, the spindle rotates 90°. Presumably astral microtubules are captured on a cortical site near TJs.

Borisy: There is a whole story that connects this with familial colon cancer and the *APC* gene.

Vallee: It is known that cytoplasmic dynein forms a belt in MDCK cells in particular in the region of the TJs. If you watch the movement of the spindle, it ends up in parallel to the cytoplasmic dynein belt. Results from our own lab indicate that LIS1 dominant-negative overexpression disrupts the belt and spindle orientation is randomized. There are these connections.

Mostov: My hunch is that this is the LIS1 and dynein systems are the target of what happens with HGF to cause a change in spindle orientation.

Vallee: It looks like in the cyst spindle orientation is normal—it is similar to what is seen in a monolayer. One interesting thing is what happens to this whole mechanism in the cells as they move out to form the tubule. What is the spindle orientation in the cells as they move out? Is it randomized, or are the spindles oriented along the axis of the emerging tubule?

Mostov: The cells look like they have spindles in the axis of the emerging tubule. It's a new monolayer, effectively.

Vallee: What is the location of the various known junctional markers?

Mostov: We are working on this.

Peter: Do you think that spindle orientation occurs, or not? It could be that the spindle is random and then the polarity is corrected after the division.

Mostov: In a monolayer not treated with HGF, they rotate 90°, stay there, and division happens. This is the Eric Karsenti mechanism. Here, we find variable degrees of this see-saw motion.

Peter: It could be that the spindle doesn't matter, but after division polarity is corrected by localization of polarity markers.

Mostov: One of the offsprings of this division loses all polarity markers as it moves out of the monolayer. This is what we call this a partial epithelial–mesenchymal transition. At this point we have correlative evidence only.

Borisy: Kozo Kaibuchi has presented evidence that APC helps cortical capture of microtubules, and if there was a way of localizing APC to other elements at the apical surface in the flanking epithelial cells, this would provide a natural mechanism for capturing the astral microtubules of the spindle and thereby orienting the axis of the spindle, and achieving division in the plane of the epithelial layer.

Nelson: There is a paper from Bienz (Hamada & Bienz 2002) who has shown in *Drosophila* that there is an APC variant that doesn't bind MTs but is localized to the adherens junction. She has implicated this in regulating that process.

Gundersen: Is cell division initiating the formation of these tubules? Do you need cell division other than to populate the extension of the tubule?

Mostov: The mitomycin C result where there are no cells migrating out shows that cell division is necessary, but not why. The Raf overexpressing cells divide less. The tubules are shorter.

References

Audebert S, Navarro C, Nourry C et al 2004 Mammalian Scribble forms a tight complex with the betaPIX exchange factor. Curr Biol 14:987–995

Coucouvanis E, Martin GR 1995 Signals for death and survival: a two-step mechanism for cavitation in the vertebrate embryo. Cell 83:279–287

Hamada F, Bienz M 2002 A Drosophila APC tumour suppressor homologue functions in cellular adhesion. Nat Cell Biol 4:208–213

Lin HH, Yang TP, Jiang ST, Yang HY, Tang MJ 1999 Bcl-2 overexpression prevents apoptosis-induced Madin-Darby canine kidney simple epithelial cyst formation. Kidney Int 55:168–178

Troxell ML, Gopalakrishnan S, McCormack J et al 2000 Inhibiting cadherin function by dominant mutant E-cadherin expression increases the extent of tight junction assembly. J Cell Sci 113:985–996

Vega-Salas DE, Salas PJ, Rodriguez-Boulan E 1987 Modulation of the expression of an apical plasma membrane protein of Madin-Darby canine kidney epithelial cells: cell–cell interactions control the appearance of a novel intracellular storage compartment. J Cell Biol 104:1249–1259

Rho GTPase–formin pairs in cytoskeletal remodelling

Kathryn M. Eisenmann, Jun Peng, Bradley J. Wallar and Arthur S. Alberts

Laboratory of Cell Structure and Signal Integration, Van Andel Research Institute, 333 Bostwick Ave., N.E., Grand Rapids, MI 49503, USA

Abstract. Diaphanous-related formins (Drfs) are members of a conserved formin family of actin-nucleating proteins and are thought to act as Rho GTPase effectors in signal transduction pathways that govern gene expression, cytoskeletal remodelling and cell division. *In vitro* evidence suggests that the three mammalian Drf proteins—mDia1, mDia2 and mDia3—have distinct GTPase-binding specificities. However, much of our current understanding of GTPase–Drf partnerships in mammalian cell signalling is based on expression studies using Drfs missing their unique GTPase-binding domains. We have employed fluorescence resonance energy transfer (FRET) and gene targeting approaches to identify the function of different GTPase–formin pairs in cell signalling. These studies have allowed us to uncover new roles for Drf proteins in cytoskeletal remodelling and novel regulatory mechanisms whereby GTPases influence formin function. Our genetic experiments strongly suggest that Drfs cooperate with other GTPase effector proteins, including the gene product of the Wiskott–Aldrich syndrome gene, WASP, during the regulation of cell proliferation. Further, the Drf gene knockout experiments indicate that this family of formins has a role in cancer pathophysiology.

2005 Signalling networks in cell shape and motility. Wiley, Chichester (Novartis Foundation Symposium 269) p 206–222

The actin cytoskeleton is a dynamic, tightly regulated protein network that plays a critical role in mediating diverse cellular processes including cell division, migration, endocytosis, vesicle trafficking and cell shape (Ridley 2001). Actin exists in either a monomeric (G-actin) or polymeric (F-actin; also called filamentous) form. F-actin is a polymer assembled almost exclusively through the addition of actin monomers to the fast-growing 'barbed (+) end' as opposed to the slower-growing 'pointed (−) end'. Nucleation of new actin polymers is the rate-limiting step to an otherwise rapid process of polymerization (Pollard et al 2000, Zigmond 2004). While for years the basic biochemistry of actin nucleation and polymerization has been a focus of intense study, the molecular mechanisms and signal transduction networks that control actin filament assembly within cells in response to extracellular stimuli have remained less well defined.

Signal transduction and spatially controlled assembly of F-actin networks

One well-characterized actin nucleator, the Arp2/3 complex, induces the formation of branched actin filaments (Higgs & Pollard 2001). Arp2/3 works by complexing with G-actin or by binding to the side of preexisting filaments (Pollard & Borisy 2003). The complex is composed of seven actin-related proteins and other subunits and is, on its own, largely inactive. Its nucleation and filament-binding activity is tightly regulated by interactions with nucleation-promoting factors (NPFs), the most prominent being the WASP/Scar family. WASP and the related N-WASP are autoregulated molecular switches controlled serially or in parallel by yet other switch-like proteins such as Cdc42, a Rho family small GTPase. Thus, in the case of WASP and Arp2/3, it is the NPFs that integrate signals controlling growth factor-stimulated actin nucleation and branching. Cdc42-activated WASP induction of Arp2/3 activity is shown schematically in Fig. 1A.

Wiskott–Aldrich syndrome: a model genetic disease caused by dysregulated cytoskeletal remodelling

WASP is encoded by a gene that is mutated in the X-linked hereditary disease called Wiskott–Aldrich syndrome (WAS), which is characterized by thrombocytopenia, eczema, increased susceptibility to infection, lymphoma and leukaemias (Badour et al 2003). There are strong genotypic–phenotypic relationships in WAS; some mutations introduce stop codons or inactivating amino acid substitutions that disrupt its ability to nucleate and remodel actin via Arp2/3. Other mutations which don't affect WASP's ability to activate Arp2/3 appear to result in a 'gain-of-function'. These WASP mutants encode proteins that fail to maintain their autoinhibited conformation (Devriendt et al 2001); in this case, affected patients additionally develop neutropaenia, which is typically not a symptom of WAS.

Studies of haematopoietic cells from human WAS patients suggest that the biological mechanisms responsible for the physiopathology are due to a deregulation of the normal remodelling of the actin cytoskeleton in response to stimuli (Snapper & Rosen 1999). For example, WASP-deficient B lymphocytes are impaired in interleukin 4 (IL4) and CD40-dependent induction of polarization and spreading. Two mouse strains having WASP deficiencies have been created by gene knockout technology (Snapper et al 1998, Snapper & Rosen 1999, Lacout et al 2003). Hemizygous WASP-deficient animals exhibit T cell deficiencies similar to those observed in WAS patients, namely, poor T cell proliferation and secretion of cytokines in response to T cell receptor stimulation. Yet WASP knockout mice do not develop symptoms of WAS. The animals develop colitis; there is no evidence of eczema. The observations strongly

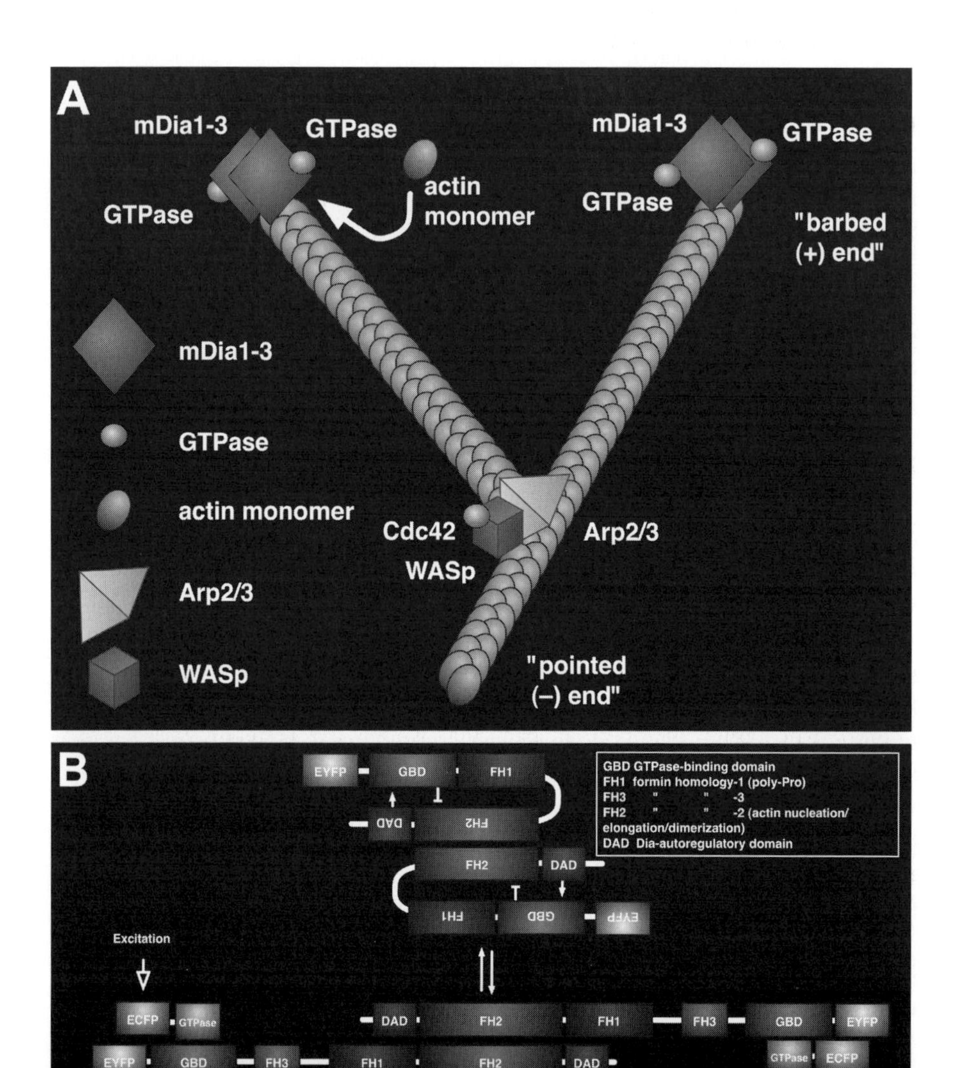
A
mDia1-3
GTPase
mDia1-3
GTPase
actin
monomer
GTPase
GTPase
"barbed
(+) end"
mDia1-3
GTPase
actin monomer
Cdc42
Arp2/3
WASp
Arp2/3
WASp
"pointed
(–) end"
B
EYFP
GBD
FH1
DAD
FH2
GBD GTPase-binding domain
FH1 formin homology-1 (poly-Pro)
FH3 " " -3
FH2 " " -2 (actin nucleation/
elongation/dimerization)
DAD Dia-autoregulatory domain
FH2
DAD
FH1
GBD
EYFP
Excitation
ECFP
GTPase
DAD
FH2
FH1
FH3
GBD
EYFP
EYFP
GBD
FH3
FH1
FH2
DAD
GTPase
ECFP
Emission (FRET)
actin monomers
(+)
"barbed-end processive
elongation"
(–)

suggest the presence of additional 'modifier' genes that influence the outcome in WAS-afflicted patients and, potentially, in model genetic systems like WASp-targeted mice (Badour et al 2003).

Rho GTPases, F-actin networks and human disease

Cdc42 is a member of the Rho family of small GTPases. These GTPases influence the organization of the actin and microtubule networks by controlling a myriad of signalling proteins (including WASP) that influence adhesion, motility, proliferation, vesicle trafficking, transcription and survival (Etienne-Manneville & Hall 2002). Rho GTPases act as molecular switches that alternate between GTP- and GDP-bound states; the activated, GTP-bound proteins preferentially interact with numerous autoregulated downstream effector proteins. Signal transduction models usually depict activated GTP-bound Rho proteins binding to and recruiting downstream effectors to their sites of action in cells. However, there is little evidence to suggest that this is the case. The GTPases themselves are more likely to be complexed with guanine-nucleotide dissociation inhibitors (GDIs) that more or less act as a reservoir within cells from which they are coaxed into action by a combination of membrane lipids and guanine nucleotide exchange factors that activate GDP release and GTP binding (Robbe et al 2003). These events likely occur at sites where various effectors are already located. Still, it appears that different GTPases control events at specific sites in cells where they fulfil their molecular-switch roles.

Some Rho family members, in particular RhoC, are overexpressed in breast cancer and malignant melanoma. This may be specific for RhoC; RhoC's ability

---

FIG. 1. Drfs are actin nucleators whose activity is regulated through interactions with small GTPases. (A) A model for formin (mDia1–3) and WASP collaborating in cells with activated Cdc42, in which Cdc42 interacts with mDia2 in cells at specific sites associated with membrane protrusions. We propose that WASP and formins collaborate to generate specific actin structures. In this model, activated WASP nucleates branched filaments from the side of mDia2 nucleated 'mother' filaments; alternatively, mDia2 binds to and processively elongates filaments after nucleation by Arp2/3. Potentially, this simple model could be expanded to include other regulators of Arp2/3, such as cortactin or Scar proteins, and other non-GTPase regulated formins. (B) GTP-bound Rho GTPase binding disrupts intramolecular interactions between the GTPase binding domain (GBD) and Diaphanous auto-regulatory domain (DAD) of a DRF. Dimerization is achieved through interactions between the FH2 domains. Shown is *intramolecular* autoinhibition. However, it is also possible that autoinhibition is achieved through *intermolecular* interactions where the GBD of one Drf interacts with the DAD of another and *vice versa*. Also, while autoinhibition can be demonstrated *in vitro*, there is no evidence that activated GTPases are sufficient to activate the autoinhibited Drfs. Whether this is due to deficiencies of the *in vitro* actin nucleation assays or to additional cooperative signals within cells has not yet been determined.

to drive invasion and metastasis cannot be matched by the closely related RhoA (van Golen et al 2002). On the other hand, RhoB has been hypothesized to be a negative regulator of growth and proliferation and is potentially a tumour suppressor protein (Prendergast 2001). In fact, RhoB expression can be lost in many types of tumours; in one recent study, more than 96% of sampled lung carcinomas had little or no RhoB protein (Mazieres et al 2004). Beyond WASP, the *in vivo* roles of the Rho GTPase effectors in the progression to malignancy are poorly understood. In particular, the roles of the Diaphanous-related formin (Drf) family of actin-nucleating proteins—which bind to both activated RhoB and RhoC *in vitro*—in growth control within normal or cancer cells have not been studied in any detail.

The Diaphanous-related formin family of Rho GTPase effector proteins

Formins are a highly conserved family of proteins implicated in a diverse array of cellular functions, most notably the cytoskeletal remodelling events necessary for cytokinesis, bud formation in yeast, establishment of cell/organelle polarity and endocytosis (reviewed in Wallar 2003, Evangelista 2003). Formin proteins were originally named for their similarity to the limb deformity gene (*ld*) originally characterized in mice (Kleinebrecht et al 1982, Castrillon & Wasserman 1994). With few exceptions, formins share two regions of homology: the FH1 domain, a proline-rich domain known to interact with SH3- and WW-domain-containing proteins, and the FH2 domain, which harbours the ability to nucleate and elongate actin filaments at the barbed end (Fig. 1A & B) (Zigmond 2004). The FH2 domain also facilitates dimerization (Wallar & Alberts 2003, Shimada et al 2004, Xu et al 2004).

The Drfs are a subfamily of formin homology proteins that, in addition to the FH1 and FH2 domains, share loosely conserved Rho GTPase-binding domains (GBDs) in their N-termini and a highly conserved Diaphanous-autoregulatory domain (DAD) in their C-termini (Alberts 2001, Wallar & Alberts 2003). The GBDs from various Drfs have been shown to interact with their DAD partner *in vitro*, leading to the autoregulation model depicted in Fig. 1B. The model shows that while the formin proteins dimerize through their FH2 domains (Shimada et al 2004, Xu et al 2004), it is the GBD–DAD interactions that act as the linchpins of autoregulation. GTP-bound Rho proteins can interact with the GBD and interrupt the autoinhibited formin, subsequently nucleating and elongating non-branched actin filaments.

There are at least two potential problems with this model. First, it is not clear whether autoregulation occurs in *cis* through intramolecular *GBD-DAD* binding or in *trans* through intermolecular binding between the dimers. For simplicity, we have drawn the *intramolecular* autoregulated configuration in Fig. 1B. Secondly, while Li & Higgs (2003) have shown that the amino terminus of mDia1 can bind to and inhibit nucleation by the isolated FH1-FH2 region of the same formin, this formin complex cannot be 'reactivated' by GTP-bound RhoA. These *in vitro* experiments suggest the possibility that there are additional signals necessary for the disruption of the autoinhibited formin.

The Drf autoregulatory mechanism can be demonstrated experimentally in cells by expressing GBD-truncated formins or by overexpression of a DAD peptide that binds to cellular Drfs and activates them by mimicking small GTPase binding (Watanabe et al 1999, Tominaga et al 2000, Alberts 2001). While much has been learned about the role of Drfs in cells by these experiments using such activated proteins, the influence of GTPases over Drf activity through direct binding is either lost or masked. It has been unclear whether GTPase binding simply activates the Drfs' ability to nucleate actin, or provides other signals that direct subcellular targeting and recruitment of Drf-associated proteins. Therefore, questions remain as to where and how Rho GTPases affect the ability of the Drfs to nucleate and processively elongate actin filaments within cells as they respond to extracellular cues or cell-cycle progression, as well as whether these mechanisms become dysregulated in disease states. Another question is, do Drfs activated in specific cellular contexts (i.e. on vesicles or at sites of adhesion) associate with or work in parallel with other modifiers of actin polymerization to generate site-specific types of F-actin networks? We have taken two approaches to address these questions in the context of mammalian cell signalling. First, using fluorescence resonance energy transfer (FRET) technology, we have exploited the knowledge that GTPases bind directly to the Drfs' N-termini to measure where and when different Rho family members interact with specific formin proteins. Second, we have knocked out *Drf* gene expression in mice.

Monitoring site-specific interactions between Rho GTPases and the mammalian diaphanous-related formins: Cdc42-mDia2 and RhoB-mDia2

Fluorescence resonance energy transfer (FRET) is a powerful technique that allows for the assay of protein–protein interactions in cells. FRET depends on the coupling of two fluorophores, in this case, cyan fluorescent protein (CFP) and yellow fluorescent protein (YFP). When fused to the GTPases and excited at the appropriate wavelength, CFP acts as a fluorescent *donor* which then excites the fluorescent *acceptor* YFP, which is fused to a Drf protein (shown schematically in Fig. 1B). FRET occurs only when the donor/acceptor pair is in close proximity, at distances (less than 30 Å) necessary for protein–protein interactions.

The CFP–GTPase and YFP–formin fusion proteins are expressed following microinjection of their respective expression plasmids into cells, which are then fixed 4 h later (Fig. 2). Cells are also counterstained with fluorescent phalloidin in order to visualize the F-actin networks. In the first example (Fig. 2A), YFP-mDia2 is expressed with CFP-Cdc42. FRET is shown in a false-temperature scale (0–16 384 arbitrary fluorescence intensity units) with a pseudocolour spectrum being applied to the image from blue (low FRET) through red (high FRET) (colours not visible in this greyscale rendering of the figures). Based on FRET signals observed, Cdc42 interacts with mDia2 primarily at two sites, at the leading cell edge (or *cortex*) and at the microtubule-organizing center (MTOC). In other experiments, we have shown that this interaction depends upon the integrity of the CRIB (Cdc42/Rac1-interaction/binding) (Burbelo et al 1995) motif within the mDia2 GBD. CRIB motifs are found in numerous Cdc42 effectors and are necessary for binding to the GTPase. The observation here indicates that one particular GTPase–formin pair, Cdc42 and mDia2, may have a role in remodelling actin at the cell edge.

What this pair contributes to actin or microtubule (MT) dynamics at the MTOC, however, remains an open question. We speculate that the Cdc42/mDia2 formin pair may be participating in the regulation of MT regulation at the minus (−) end of MTs, which, like actin, are assembled and disassembled in a polarized fashion (Gundersen 2002). We also speculate that other GTPase–formin pairs, such as RhoA and mDia1 or mDia2, may be working at the plus end of microtubules to direct them to focal adhesions or other sites. To this end, from collaborative efforts with the Gundersen lab, it has been shown that mDia1 and mDia2 can complex with the MT (+)-end binding proteins APC and EB1 (Wen et al 2004).

In contrast to the interaction between Cdc42 and mDia2 at the MTOC and the leading edge, RhoB is already known to have a role in endocytic or vesicular trafficking (Qualmann & Mellor 2003), and it interacts with mDia2 on endosomes (Fig. 2B). This result is consistent with our discovery of the localization of both mDia1 and mDia2 on endosomes (Tominaga et al 2000). The expression of either activated RhoB or deregulated versions of mDia1, mDia2 and mDia3 interferes with endosomal trafficking (Gasman et al 2003); each blocks movement of vesicles as well as increases their number. One interpretation of these observations is that expression of RhoB or deregulated mDia1–3 triggers an inappropriate transition from fast MT-dependent transport to actin-dependent transport (Randazzo 2003).

A second interpretation, depicted in Fig. 3, is based on a model proposed by Maniak and colleagues (Drengk et al 2003). In this case, formins promote the assembly of F-actin on vesicles containing endocytosed cargo; in our experiments the cargo is internalized fluorescent epidermal growth factor (TR-EGF). Disassembly of the F-actin coats leads to inappropriate fusion of vesicles, which then appear as 'grape-like clusters'. As shown by the Maniak Lab, disassembly can be achieved by treating cells with the actin-depolymerizing drug

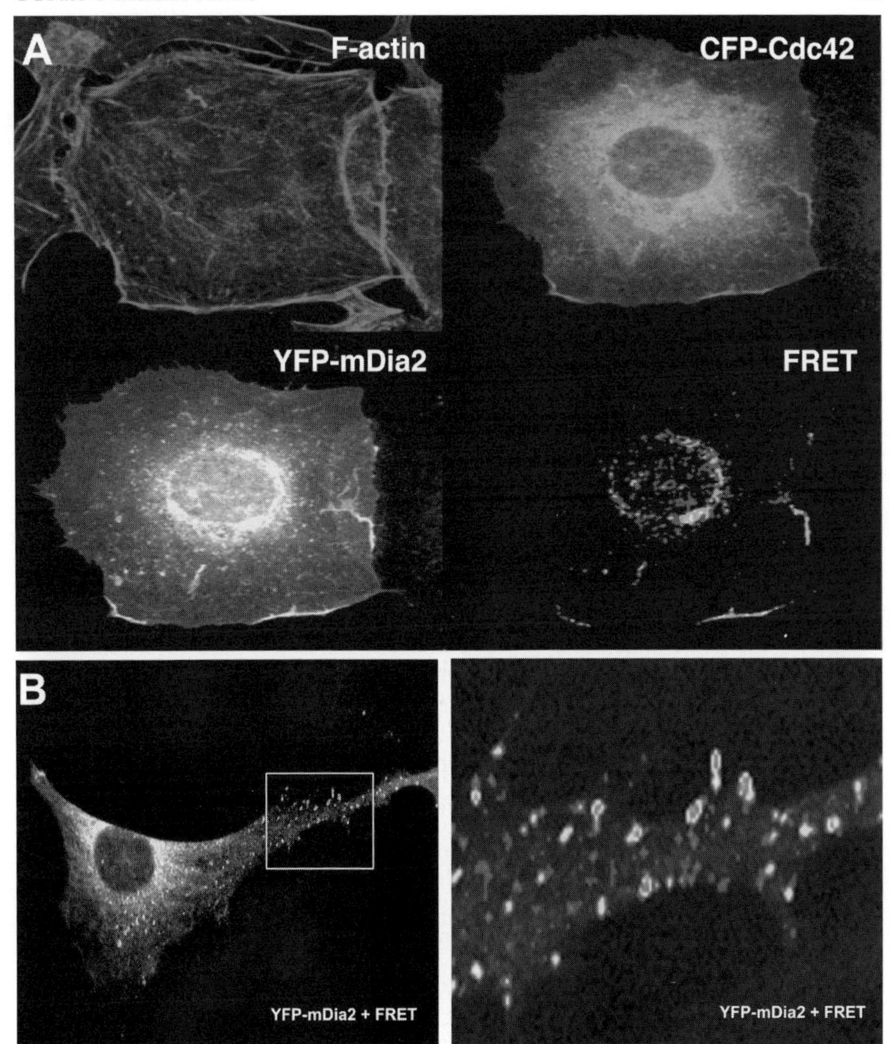

FIG. 2. Different GTPases interact with mDia2 at specific sites. (A) Cdc42 binds to mDia2 at the cell cortex and at the MTOC. CFP-fused Cdc42 and YFP-fused mDia2 were in cells following microinjection of their respective expression plasmids; FRET was assessed 4 h later. A distinct FRET signal is observed between Cdc42 and mDia2 at the MTOC, the cell cortex, and along microspikes. (B) RhoB interacts with mDia2 on vesicles. CFP-fused activated RhoB-G14V and YFP-fused mDia2 were co-injected into cells and FRET assessed 4 h later. A FRET signal is observed between activated RhoB and mDia2 upon endosomes. These data indicate that individual GTPase-formin pairs would appear to be functionally distinct, promoting the formation of different actin structures at different sites. Since both RhoB and Cdc42 have been implicated in endocytosis and vesicle trafficking, it is possible that both GTPases utilize mDia2 sequentially or in parallel as effectors to transport cargo within cells.

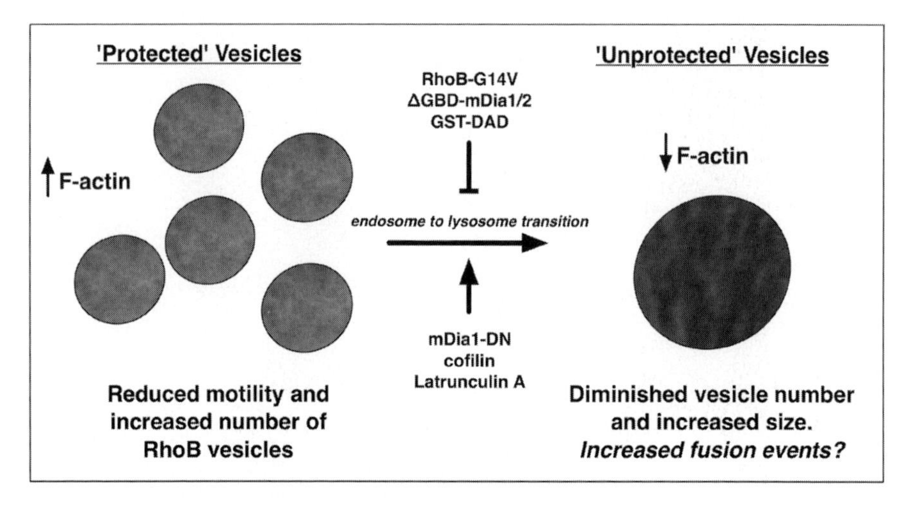

FIG. 3. A proposed role for formins in maintaining vesicular integrity through F-actin accumulation. Based on a model by Drengk et al (2003), the model displays a potential role of the RhoB–mDia2 interaction in vesicle trafficking. The RhoB–mDia2 pair assembles or maintains F-actin coats on vesicles bearing internalized cargo. The purpose of the F-actin coat is to protect the vesicle from inappropriate fusions. The expression of deregulated formins or activated RhoB prohibits changes in F-actin on vesicles that allow fusion. In contrast, expression of dominant-negative mDia1 (mDia1-DN) or the treatment of cells with the actin-depolymerizing drug cytochalasin D reverses this process, leading to an increase in multi-vesicular bodies and the appearance of grape-like clusters caused by improper vesicular fusion.

Latrunculin A or by expression of an endosomally targeted version of the actin-severing protein cofilin (Drengk et al 2003). In our experiments, we can replicate these effects by expression of a dominant-interfering version of mDia1 (mDia1-DN) (Copeland & Treisman 2002). Taken together, these experiments point to a new role for formins in actin remodelling, whereby they control not only the assembly of actin-myosin conduits for vesicle trafficking, but also the F-actin on

---

FIG. 4. *Drf1* gene targeting leads to a susceptibility to malignancies and suggests a role for Drfs in T cell signalling. (A&B) H&E staining of liver (A) and lung (B) of a 9-month-old *Drf1*$^{-/-}$ mouse with tumour; insets show that small cell lymphoma had infiltrated both sites. (C) FRET assays show that both RhoA and Cdc42 interact with formins upon activation of the T cell receptor. Human leukaemic Jurkat T cells expressing YFP-fused mDia1 and CFP-fused RhoA (top), YFP-fused mDia1 and CFP-fused Cdc42 (middle), and YFP-fused mDia2 and CFP-fused Cdc42 (bottom) were activated by TCR ligation for 10 min to anti-CD3-coated glass coverslips and FRET analysed in fixed cells. A FRET signal is observed between mDia1 and RhoA upon T cell activation. A FRET signal is observed between mDia2 and Cdc42, yet not Cdc42 and mDia1, which lacks the conserved CRIB domain found in numerous Cdc42 effector proteins.

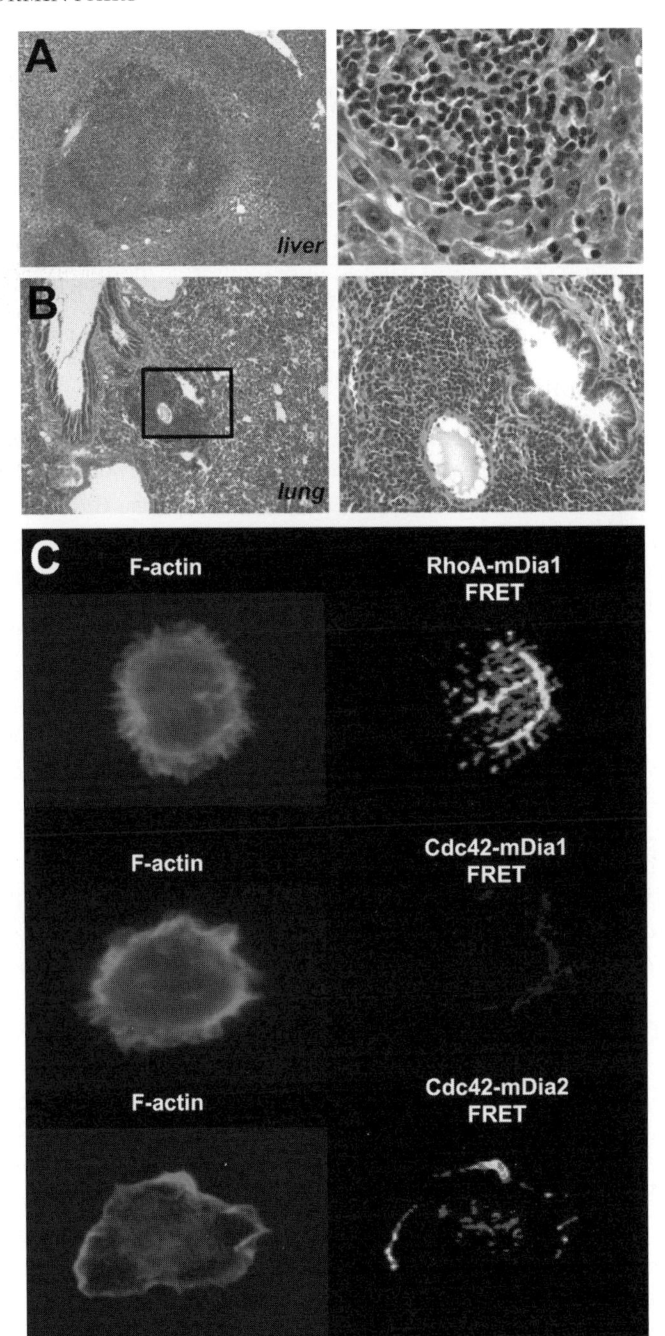
A
liver
B
lung
C
F-actin
RhoA-mDia1
FRET
F-actin
Cdc42-mDia1
FRET
F-actin
Cdc42-mDia2
FRET

vesicles that organize intracellular cargo. This model implies that different Rho GTPase–formin pairs should have critical roles in trafficking and the regulation of growth factor receptors. Disruption of the normal formin function in this process should have significant effects on the proliferation of not only normal but also tumour cells. Further, these observations raise the question of whether or not formins have a role in the cancer pathophysiology. Our genetic studies of the gene encoding mDia1 (*Drf1*) in mice suggest that formins not only have a role in the development of cancer, but also have an underlying role in immune cell function that is phenotypically similar to that of WAS.

WASP–formin collaboration

We have taken genetic approach to elucidate the role of the Drf proteins *in vivo*. Like WAS patients, who have an increased incidence of malignancies, *mDia1* knockout (KO) mice show, in addition to lymphodysplasia, the development of cancer as they approach 7–10 months of age. The animals also develop splenomegaly, lymphopenia, thrombocytopenia and neutropenia. The cancers observed in *mDia1* KO mice include lymphoma (as shown in Fig. 4A), histiocytic sarcoma and adenocarcinoma. Concurrent with the development of disease, we have found that T cell responses to proliferative stimuli are amplified. However, in younger animals—again like WAS KO mice or human WAS—T cell proliferative responses to activation of the T cell receptor/CD3 (TCR) complex is defective. Because of these observations, our data suggests that mDia1 is a critical component of T cell responses. We also hypothesize that in place of mDia1 loss, mDia2 or mDia3 up-regulation accounts for the hyperreactivity seen in older animals. The up-regulation of mDia2 or mDia3 expression may then be the underlying cause of the cancer pathophysiology. Our ongoing experiments are testing these related hypotheses.

The defects in T cell signalling prompted us to investigate the role of Drf proteins in T cell signalling to the cytoskeleton in cells. T cell activation by CD3 ligation of the TCR induces rapid and dramatic cytoskeletal reorganization. Consistent with the effects of the ablation of the mDia1 gene, expression of the dominant interfering mDia1 blocked TCR-activated actin remodelling in transfected Jurkat T cells. To determine which GTPase might regulate mDia1 activity in activated T cells, we turned to FRET-based technology. Different CFP–GTPase fusion proteins were transfected into Jurkat cells, which were then stimulated with anti-TCR antibodies. While Cdc42 interacts with mDia2 (Fig. 4B; bottom), as previously observed in fibroblasts (Fig. 2B), it does not bind to mDia1. Instead, RhoA binds to mDia1 as cells react to TCR activation (Fig. 2B, top). These experiments point to an important role for mDia1 and mDia2 as effectors for small GTPases in mediating proliferative responses in activated T cells.

Conclusions

Our ongoing studies of Rho GTPase–formin pairs show that the formins can act as actin assembly machines at different sites within cells associated with very different types of F-actin networks. Functioning as an effector for Cdc42, mDia2 participates in remodelling of cortical actin associated with leading-edge dynamics. In contrast, our data suggest that RhoB participates in vesicular trafficking, not only by mediating the production of actin filaments or microtubules necessary for motor-dependent trafficking, but also through the assembly of F-actin upon vesicles, which is required for their integrity. These studies imply important roles for formins in the control of growth factor receptor dynamics within cells. Lastly, through genetic studies of the mouse *Drf1* gene that encodes mDia1, we find that mDia1 (and potentially mDia2) are critical components of T cell responses to proliferative stimuli. These important studies suggest that defective human *DRF* genes may play a role in cancer pathophysiology.

Acknowledgements

ASA was supported by Department of Defense Breast Cancer Research Program Career Development Award (DAMD17-00-1-0190) and the Van Andel Foundation; additional support to ASA and JP were provided by the National Cancer Institute (R21 CA107529). KME was supported by Ruth L. Kirschstein National Research Service Award (F32 GM072331) from the National Institute of General Medical Sciences. We are grateful to David Nadziejka, Nick Duesbery, Harry Higgs, Gregg Gundersen and Kathy Siminovitch for comments and numerous discussions.

References

Alberts AS 2001 Identification of a carboxy-terminal diaphanous-related formin homology protein autoregulatory domain. J Biol Chem 276:2824–2830

Badour K, Zhang J, Siminovitch KA 2003 The Wiskott-Aldrich syndrome protein: forging the link between actin and cell activation. Immunol Rev 192:98–112

Burbelo PD, Drechsel D, Hall A 1995 A conserved binding motif defines numerous candidate target proteins for both Cdc42 and Rac GTPases. J Biol Chem 270:29071–29074

Castrillon DH, Wasserman SA 1994 Diaphanous is required for cytokinesis in Drosophila and shares domains of similarity with the products of the limb deformity gene. Development 120:3367–3377

Copeland JW, Treisman R 2002 The Diaphanous-related Formin mDia1 Controls Serum Response Factor Activity through its Effects on Actin Polymerization. Mol Biol Cell 13:4088–4099

Devriendt K, Kim AS, Mathijs G et al 2001 Constitutively activating mutation in WASP causes X-linked severe congenital neutropenia. Nat Genet 27:313–317

Drengk A, Fritsch J, Schmauch C, Ruhling H, Maniak M 2003 A coat of filamentous actin prevents clustering of late-endosomal vacuoles in vivo. Curr Biol 13:1814–1819

Etienne-Manneville S, Hall A 2002 Rho GTPases in cell biology. Nature 420:629–635

Gasman S, Kalaidzidis Y, Zerial M 2003 RhoD regulates endosome dynamics through Diaphanous-related Formin and Src tyrosine kinase. Nat Cell Biol 5:195–204 (Erratum in: Nat Cell Biol 5:680)

Gundersen GG 2002 Evolutionary conservation of microtubule-capture mechanisms. Nat Rev Mol Cell Biol 3:296–304

Higgs HN, Pollard TD 2001 Regulation of actin filament network formation through ARP2/3 complex: activation by a diverse array of proteins. Annu Rev Biochem 70: 649–676

Kleinebrecht J, Selow J, Winkler W 1982 The mouse mutant limb-deformity (ld). Anat Anz 152:313–324

Lacout C, Haddad E, Sabri S et al 2003 A defect in hematopoietic stem cell migration explains the nonrandom X-chromosome inactivation in carriers of Wiskott-Aldrich syndrome. Blood 102:1282–1289

Li F, Higgs HN 2003 The mouse Formin mDia1 is a potent actin nucleation factor regulated by autoinhibition. Curr Biol 13:1335–1340

Mazieres J, Antonia T, Daste G et al 2004 Loss of RhoB expression in human lung cancer progression. Clin Cancer Res 10:2742–2750

Pollard TD, Borisy GG 2003 Cellular motility driven by assembly and disassembly of actin filaments. Cell 112:453–465

Pollard TD, Blanchoin L, Mullins RD 2000 Molecular mechanisms controlling actin filament dynamics in nonmuscle cells. Annu Rev Biophys Biomol Struct 29:545–576

Prendergast GC 2001 Actin' up: RhoB in cancer and apoptosis. Nat Rev Cancer 1:162–168

Qualmann B, Mellor H 2003 Regulation of endocytic traffic by Rho GTPases. Biochem J 371:233–241

Randazzo PA 2003 RhoD, Src, and hDia2C in endosome motility. Dev Cell 4:287–288

Ridley AJ 2001 Rho proteins: linking signaling with membrane trafficking. Traffic 2:303–310

Robbe K, Otto-Bruc A, Chardin P, Antonny B 2003 Dissociation of GDP dissociation inhibitor and membrane translocation are required for efficient activation of Rac by the Dbl homology-pleckstrin homology region of Tiam. J Biol Chem 278:4756–4762

Shimada A, Nyitrai M, Vetter IR et al 2004 The core FH2 domain of diaphanous-related formins is an elongated actin binding protein that inhibits polymerization. Mol Cell 13: 511–522

Snapper SB, Rosen FS 1999 The Wiskott-Aldrich syndrome protein (WASP): roles in signaling and cytoskeletal organization. Annu Rev Immunol 17:905–929

Snapper SB, Rosen FS, Mizoguchi E et al 1998 Wiskott-Aldrich syndrome protein-deficient mice reveal a role for WASP in T but not B cell activation. Immunity 9:81–91

Tominaga T, Sahai E, Chardin P et al 2000 Diaphanous-related formins bridge Rho GTPase and Src tyrosine kinase signaling. Mol Cell 5:13–25

van Golen KL, Bao LW, Pan Q et al 2002 Mitogen activated protein kinase pathway is involved in RhoC GTPase induced motility, invasion and angiogenesis in inflammatory breast cancer. Clin Exp Metastasis 19:301–311

Wallar BJ, Alberts AS 2003 The formins: active scaffolds that remodel the cytoskeleton. Trends Cell Biol 13:435–446

Watanabe N, Kato T, Fujita A, Ishizaki T, Narumiya S 1999 Cooperation between mDia1 and ROCK in Rho-induced actin reorganization. Nat Cell Biol 1:136–143

Wen YI, Eng CH, Schmoranzer J et al 2004 EB1 and APC interact with mDia to selectively stabilize microtubules downstream of Rho and promote cell migration. Nat Cell Biol 6:820–830

Xu Y, Moseley JB, Sagot I et al 2004 Crystal structures of a formin homology-2 domain reveal a tethered dimer architecture. Cell 116:711–723

Zigmond SH 2004 Formin-induced nucleation of actin filaments. Curr Opin Cell Biol 16: 99–105

DISCUSSION

Yap: What does the bone marrow of your mice look like?

Alberts: It depends on the strain and the phenotypes are complex. We see various myelodysplasias at different ages. Thus, it doesn't always look like leukaemia, but there are clearly problems at all ages and genetic backgrounds.

Yap: It looks like human myelodysplasia, which often presents with lineage deficiencies and you get a variety of haemopoetic malignancies later.

Alberts: That fits with the 5q minus syndrome (loss of 5q31). Now we have made these observations we are going to go back to look at gene expression changes by RT-PCR, tracking *mDia1*, *mDia2* and *mDia3*.

Nelson: Have you looked specifically at different lineages within the bone marrow?

Alberts: No, but we are in collaboration with Kathy Siminovitch, an expert in the field who is helping us along. She has our mice and we have some of hers. She has WASP knockout mice and we intend on pursuing breeding experiments.

Luca: You mentioned the testes as having cytokinesis defects. Do you see that in a lot of other tissues?

Alberts: This is the only tissue we have seen it in. I'd emphasize that we only saw that in heterozygous tissues. We haven't looked in ovaries. In the original Wasserman paper on Diaphanous they saw similar defects in ovaries in addition to testes. This would be the best candidate tissue to look in.

Luca: Do you think it is meiotic specific?

Alberts: I don't know. One of my colleagues is a testis/germ cell expert. One possibility is that there are problems in the way that syncytia are formed. Typically these cells stay in one syncytium of around 200 cells. There may be some problem with this. It looks like they are not getting past the secondary spermatocyte step. He suggested we do things as involved as injecting the animals' tissue wash with propidium iodide and watch meiotic divisions.

Kaibuchi: In terms of a FRET signal, could you see some active form of mDia in the periphery?

Alberts: I didn't show you data in fibroblasts with RhoC. We see Cdc42 and mDia2 interacting at the cell edge. We see also RhoC and mDia1 interacting clearly at the lamella in migrating in several cell types. It is hard to do these kinds of experiments. We think that there is some occurring at the edge and some occurring further back on other as yet unidentified structures.

Peter: You also said you see interactions with endosomes.

Alberts: We see RhoB interacting on endosomes. These endosomes are defined by anti-Lamp1 and EEA1 staining and contain Texas red-labelled epidermal growth factor (EGF). Extensive work by Harry Mellor has shown that RhoB can be a marker of early to late endosome transitions.

Peter: Does trafficking affect the function of formins?

Alberts: Formin moves with them. As with activated mDia and RhoD, if we express activated versions of RhoB, we impede trafficking. It is not clear, however, what actin structures are being generated on the endosomes. We have become more interested in the potential role of these things in the downstream signalling from EGF.

Chang: I have a question about possible connections between formins and the Arp2/3 complex. Can you immunoprecipitate formin with Arp2/3 proteins?

Alberts: Yes. It has been frustrating because we don't understand what this might imply. With Mike Rosen's lab we took DAD, the autoregulatory domain, which looks very much like a WH2 domain and tested the hypothesis that it might activate Arp2/3. We found that it didn't. While we were working on this idea, the Bretscher, Boone and Pellman labs, working on Bni1, came up with the observation that the FH2 domain could nucleate actin. Formins can complex with Arp2/3 and they colocalize nicely if you stimulate cells with Cdc42, but we don't think it is direct. These observations have led to our working hypothesis that formins and WASP-like molecules might collaborate to generate specific actin structures.

Borisy: In flies that are deficient in WASP function, the phenotype is primarily expressed in the nervous system. Much of the fly is normal: it is certainly viable. In this kind of a situation, how would you see mDia function operating? If mDia doesn't have a WASP partner yet cells are forming normally in most of the fly, would it be interacting with Scar/Wave, or might it be operating independently?

Alberts: The Wave2 phenotype is very similar to the mDia1 knockout phenotype. In the absence of WASP the formins might be able to collaborate with other Wasp/Scar family members. I see no reason to eliminate this possibility.

Borisy: If the Wave2 phenotype is similar to the mDia phenotype, then this suggests that the primary partner is not WASP but Wave.

Alberts: One of many potential partners. We have two formins that are activated by Cdc42. I can imagine a situation where you have an exchange factor that turns on Cdc42 and it stimulates both WASP and the formin at the same place at the same time. The question is, which one comes first? Does the formin jump onto a newly generated filament, nucleated by WASP and Arp2/3, or the other way round? I would favour the idea that since WASP and Arp2/3 prefer binding to freshly minted filaments, the formin nucleates the filament and then Wasp and Arp2/3 jump on to the site of that filament to nucleate it.

Chang: In pombe, Kathy Gould showed that the formin was bridged to myosin type 1. This is another activator of Arp2/3, at least in fungi, through a protein called Cdc15. Have you looked at this?

Alberts: Myosin II can be coimmunoprecipitated with mDia2. This is all we know. Also, in filopodia, the formin zips back down the spines of the filopodia, so I suspect that might be something that is important. We haven't looked at myosin I yet.

Chang: Have you looked at the Cdc15 homologues, such as Pstp ip?

Alberts: Good question. This is in a complex with WASP. There is a functional interaction there.

Drubin: What are your thoughts on other functions of formins that are independent of actin assembly? In yeast we were surprised to find that actin function is not the most important part of the function of WASP. Scaffolding seems to be important.

Alberts: We have done an extensive amount of work on this. Before the FH2 domain was discovered to nucleate actin, our sole guiding hypothesis was that formins are scaffolds, much like WASP. This is why we went down the road of testing whether it activated Arp2/3. We also know that they bind Src, though the importance of that interaction has been fully investigated. In my lab, at least, we only have one protein that interacts with the formins in a manner that is regulated by GTPase binding, that is EF-1α. The interaction was first discovered for Bni1p by the Takai lab. Condeelis has shown that EF-1α, in addition to its role as a translation factor, is an F-actin binding protein. This might be another one of those collaborating actin-binding proteins that may be involved with formins in cells.

Drubin: Might there be phenotypes that are due to other interactions with GTPases?

Alberts: We have generated a host of different mutants. It is clear that formins also function as scaffolds.

Gundersen: Do you have the *mDia1* knockout T cells?

Alberts: We are pulling them out now. There is a lot to do.

Gundersen: I was wondering whether they oriented their MTOC normally.

Alberts: That is something we want to examine. We have to pull out the cells and incubate them with APCs, to see whether they form immune synapses and then orient their microtubules towards the immune synapse correctly.

Schejter: Is it clear that the GTPases that bind to formins bind in a GTP-bound state? I thought this was controversial.

Alberts: In our hands, in *in vitro* experiments, yes. They bind in a GTP-dependent manner. Higgs has done experiments slightly differently and has shown that there is some binding to GDP-bound RhoA with mDia1. Interestingly, when Eric Sahai generated mutations in the GTPase that would block binding with the formin, the only disruptive mutations were ones that disrupted the ability of the GTPase to bind nucleotide.

Gundersen: It is also worth mentioning that of the different Rho effectors, mDia binds Rho much tighter than some of the other effectors.

Alberts: From comments by Fred Wittinghofer at the FASEB meeting, there appears to be a structure coming along between RhoC and mDia1 that might tell us a lot more about how this interaction occurs. I think it will be unique.

Drubin: Did the mutant mice have any auditory problems?

Alberts: No. We have looked at this.

Final general discussion

Sheetz: Over the last few days I've been aware that with most of the phenomena we have been discussing, we have the ability to define the various steps involved. We have seen a number of studies, particularly in yeast, where it has been possible to follow kinetically each step along the way. The dynamics in most of these systems is such that if we go by a beginning point now and an endpoint even 10 minutes from now, a lot will have happened in that period that we won't understand. My point is, it is necessary for us to understand these molecules and to define the various steps in every function. We often use the analogy of the Martian coming down and trying to understand how an automobile works. If we think about the overall process of getting the car from point A to point B, any one of many hundreds of different things can go wrong. In terms of troubleshooting, an understanding of the individual steps is critical. Until we start defining the actual steps that underlie these processes on a second-by-second basis, we can't define the roles of various proteins. David Drubin's study is a particularly interesting one, looking at the formation of these endosomal structures (Drubin et al 2005, this volume). There we have some very good dynamics and we can follow this process in intimate detail. There are about 10 or 20 different steps for which we could come up with kinetic concepts. This is a very simple process to begin with. In all these other processes we have the ability to see something happening and look at the steps involved. But until we quantify these steps it is hard to come up with a true understanding of what underlies a defect.

Humphries: So what if there are 200 components, or more? Is suspect this will be the case. For example, in Benny Geiger's cartoon of what is in a focal adhesion there are dozens of components and everything is interacting with everything.

Sheetz: We have heard many situations where the overall phenomenon is altered. The question then becomes at what step in this overall phenomenon is the alteration occurring? Can it be broken down functionally to a given step?

Humphries: Is there a subset of those components that are critical? I guess this is what you are saying.

Sheetz: In the best case scenario we would find 10 or 20 components which give the same phenotypic alteration of a given step.

Borisy: Please define a step!

Sheetz: A step for me, in looking at fibroblast spreading for example, would be the extension of a lamellipodium.

Borisy: I would call that a behaviour, not a molecular step.

Sheetz: Yes, but we can quantify it; it has a beginning and an end.

Borisy: So you are not calling for identifying molecular steps. You wish, however, to define steps in a process for the purpose of analysis.

Sheetz: I am looking for something that can be defined on a second-by-second timescale. We want definitions of the rates.

Borisy: It has to have defined endpoints and be quantifiable.

Sheetz: That's right.

Chang: This talk about individual steps reminds me of some discussions in the cell cycle field, perhaps 20 years ago. It was thought that the cell cycle might be composed of discrete events that are linked together. We now know that this is not the case. There is much more complex regulation, with oscillatory elements and so on, which are talking to each other but which may be independent.

Borisy: If we are not going to think about linear processes and discrete steps, how should we think? You stated the problem; what's the solution?

Chang: We need to be open to the possibility that when we knockout something it doesn't just affect the step and everything downstream of it. These pathways may not be so linear.

Borisy: Non-linearity has come up in some of the discussions. We are dealing with complex processes. In many cases behaviour arises out of these complex systems in ways that are not obvious from analysis of any of the subsytems or components. In general, such behaviours are called emergent phenomena. What will be required is some kind of systems analysis of the complexity of the interactions in order to understand the behaviour.

Nelson: I'd like to mention the interactions maps. My problem with them is that they are pooling data from many different conditions by different people using different reagents and different criteria for specificity or not. An interaction map where A is bound to C through 24 other proteins doesn't mean that this has been demonstrated as a complex of proteins. What people have shown is that A binds to B, they may have shown that B binds to C, but they haven't shown that A binds to C through B. This is one of the major problems we are coming to. The 'new biochemistry' that is called proteomics is trying to define complicated structures. The genetic epistasis maps at least allowed one to order a certain series of interactions and allowed us to look at interactions between the genes in a pathway. These have defined some of the major interactions involved. One of the things that biochemists and cell biologists now have to do is to think carefully about the protein interactions. There are many of these models around, and I am not sure how interesting they really are. They just tell us that it is a very complex process.

Borisy: James offers us another cautionary note. You are expanding on Fred Chang's theme and saying not only beware of linear explanations, but also beware of interaction maps. That's another problem. Can you offer a solution?

Nelson: The solution is that you take one point and then work away from it. This is what geneticists have done. They have taken a specific gene and then worked upwards, forwards and backwards from this.

Humphries: To some extent you may be right. I heard Benny Geiger speak a couple of months ago. He has done multi-colour analysis of components pixel-by-pixel, almost like an imaging microarray experiment. He sees different flavours of complexes. I suspect that there are lots of teeming soups of molecules that are interacting together. The challenge is how we deconvolute this to get a meaningful signal. At the end of the day the cell does divide and a different adhesion does form in a cell, so this system must actually create a read-out.

Firtel: Whether or not we use proteomics, we still have to get back to demonstrating that an interaction has an *in vivo* function. The best way of doing this is through genetics. I'm not just talking of knockouts, because null mutations can have very abrupt effects. If you think that two molecules are interacting in a particular mechanism, then point mutations that can affect that interaction need to be done. Now with RNAi and replacement with a construct that is not going to be inhibited by that RNAi it is possible to look at direct interactions and ask whether we need these interactions in order to have a functional complex. It is one thing to describe the protein–protein interactions, and these are essential, but we need to demonstrate that these things are biologically functional and what the mechanism is.

Sheetz: That is my point, about functional complexes as opposed to lines on a diagram.

D Lane: There are two tools that are increasingly useful in this area. Both address the question that Rick Firtel has raised, which is being able to go into the system with a disruptor. One is that in many cases it is possible to define interactions down to a small peptide, and this can be presented on a protein scaffold. In my own field this has had a huge effect: this was how we could show that in every cell Mdm2 was continually regulating p53. Before this, we were overexpressing one component or another. By putting in this disruptor we could show that this was a constitutive pathway working all the time. The other thing that should be used much more is the technology of dimerizing drugs. Two proteins can be brought together using a small molecule. It is possible to split up a protein into two domains and then ask whether it matters that these two domains are together. This technique could answer the fibronectin question we were discussing.

Alberts: One added level of complexity is that these maps are often drawn in a linear fashion where one protein interacts with another. One spectre that hasn't been discussed is allosteric interactions: two proteins binding the same protein affects the interaction with a third protein, or its activity. This is often not considered.

D Lane: It has become a real issue in comparing drug inactivation of a target compared with knockouts. In many cases the drug inactivation results in the stable presence of a ligand–protein complex that can go off and do all sorts of other things. It is not reasonable to equate that behaviour with the absence of that species. Many proteins now appear to have multiple interactions and these interactions are changed by their binding partners. This can't be mimicked in a knockout.

Sheetz: The other component that hasn't been discussed is force. There is quite a bit of evidence to say that fibronectin unfolds, and this unfolding in response to force changes the epitopes. We are finding evidence inside cells of a similar type of change, which activates the Src family kinases. They appear to be under the control of mechanical force and are activated by mechanical force. Even though we know about the interactions *in vitro*, we cannot generally apply force to molecules *in vitro*. *In vivo* they are situated where the cell applies force.

Borisy: Mike points out that an underdeveloped area of cell biology is the dependence of reactions on force. I agree, completely. This applies particularly in our field where we are talking about formation of shapes and moving against resistance. What does it mean to take force into account? It means that there will be chemical reactions that are force dependent. Progress in this area may require generating strain sensors and doing kinetics and analysing steps not just in isolation but under a load.

Humphries: I suspect that the antibody I spoke about is actually detecting stretched integrin.

Chang: Are there good assays to detect strain?

Borisy: Some are under development. For example, the filamin molecule has a little coiled coil structure. You can put in a cyan fluorescent protein (CFP) and yellow fluorescent protein (YFP) to each element of the dimer and when they are close together you have a fluorescence resonance energy transfer (FRET) signal. If force is applied and they are pulled apart then the FRET signal diminishes.

Sheetz: At a functional level if we stretch cytoskeletons we activate a variety of G protein pathways.

D Lane: One topic I haven't heard discussed here is the emerging link between transcriptional regulation, actin and movement. Are there other examples of this link between transcription and cytoskeleton regulation?

Alberts: Richard Treisman presented this topic at the FASEB small G protein meeting a month ago. He has presented further evidence that a transcriptional coactivator that binds to the transcription factor SRF does indeed bind G-actin. There is also a component for MAPK signalling in this signalling mechanism.

Interestingly, the coactivator interacts through the same domain that Elk uses for SRF interactions. Elk interacts with SRF. It is the target for growth factor-controlled MAPK regulation of c-Fos whose promoter is centrally regulated by SRF. So, there is some overlap between the G-actin sensing pathway and MAPK signalling. This is potentially important for growth control and disease because the coactivator for actin is translocated in certain leukaemias.

Borisy: No one has brought up the need for a systems approach involving computational modelling. Is this too complex?

Kaibuchi: In collaboration with Dr Shinya Kuroda at University of Tokyo, we have been trying to use a systems approach. We would like to know how myosin phosphorylation is regulated by extracellular signals. There are some big unresolved questions. Phosphorylation of myosin light chain usually continues for a long time, like 60 min, especially in smooth muscle cells. Theoretically, myosin phosphorylation can be induced for just 1–3 min, because myosin light chain kinase is activated for just 1–3 min. Rho activation continues for just 10 min at most. However, Rho kinase activation continues for longer than an hour. According to the period of Rho activation, we expect that Rho kinase activation continues for 10 min at most. We can predict that there should be some positive feedback from Rho-kinase, which can be predicted by the simulation. We use 20–30 equations to simulate signalling molecules including receptor, heterotrimeric G proteins, Rho, GEF and myosin light chain kinase, and myosin phosphatase. There are two big advantages when we use the systems approach. One is that we can predict the missing pathways, because we cannot adopt the simulation curve to the experimental results without them. Another is that we can inactivate or activate a single pathway or many pathways simultaneously in a computer. This is useful to identify the therapeutic target. We can predict what process should be inactivated to suppress hypertension or vasospasms. The answer is very simple: Rho kinase is the best target.

Borisy: These are analytical equations, but you do numerical calculations.

Sheetz: The difficulty with these is that unless you have the ability to measure things, once you have more than five adjustable parameters you can come up with any solution you wish.

Kaibuchi: Yes, that's true. We need to measure a lot of things.

Borisy: What Mike is expressing is a general distrust of models and modellers. In general, this approach has a bad aroma associated with it, because it is so flexible when there are several variables. On the other hand, if we are dealing with complex systems that have to be deconstructed and put back together, is there an alternative? At some point aren't we going to have to use computational approaches.

Sheetz: The point is that we have phenomena that we can quantify well enough that we can probably now make useful models.

Borisy: We haven't heard any modelling presentations at this meeting, but we will at the next, I suspect. What can I say in closing? I began by asking the general question about how cells become polarized. How do they break symmetry? How do they maintain polarity? How do they sense direction? Let's take a brief look at some of the themes that have emerged during the course of this meeting. In thinking about the generation of form, where does the information come from? You can imagine that it comes from a preexisting piece of information, or that there is a need to generate information. There could be a preexisting axis. In the case of *Schizosaccharomyces pombe* there is a preexisting axis the length of the cells, and in *Saccharomyces cerevisiae* there is the preexisting bud scar, although we don't know how this helps to form the site of the next bud. But any two points define an axis or line. It could be a bud scar and the nucleus, or one edge of a cell and another edge of a cell. Also, the normal to the plane defines a line. The epithelial cell contacting the substratum has a cue about where to grow. In yeast or the epithelium in contact with the substratum there is already information about where to go, and we are trying to understand how to read out this information. Then we have other cases where there is a need to generate an axis, where symmetry needs to be broken. We see this at different levels of biological organization. At the supramolecular level we know that beads with Arp activators in the right mix of proteins can make a cloud of actin filaments which then spontaneously convert into a polarized comet tail. This is a molecular system with a built-in capacity to become polarized, just with some fluctuation. Cells also display this. Cells can spontaneously break symmetry and become polarized. Keith Mostov has talked about spherical balls of cells at the tissue level which can break symmetry and develop a luminal and apical surface. So at several different levels biological systems have the capacity to break symmetry and generate an axis. Another theme that has emerged in the meeting is the distinction between polarization of the cell and orientation. We have seen this in *Dictyostelium* with mutations that allow cells to orient or move chemotactically even though the cell is not polarized, and vice versa. This is also seen in mating yeast where they may polarize but fail to orient towards each other. It is also seen in epithelial morphogenesis. A common theme is therefore that cells have some capacity for polarizing but then there is another level we have to consider, namely the orientation of that polarity axis. What are the elements that go into reading out this positional information, either preexisting or generated from breaking of symmetry? There are many. Some of the presentations have focused on small GTPases and the other molecules of the signalling swamp. The scaffolding proteins interact with the elements that generate form. The plus end binding proteins of microtubules are attracting a lot of attention. Actin filaments and their barbed end binding proteins are also a focus. There is now a lot of interest in the cross-talk molecules such as IQGAP that link the microtubules and the actin together, and also the molecular motors that interact with these polar dynamic polymers to generate form. Not only are the internal elements critical but also the external contact elements are significant. These are just some of the categories of elements whose interaction we have to understand. Then there are critical processes by which these elements interact with each other. There is a need for amplification with regard to orientation of cells responding to attractants which involves various feedback mechanisms which are imperfectly understood. This theme of feedback loops has emerged in a number of contexts throughout the meeting. Then we touched briefly on what we need in order to understand these systems, and how shape and motility arise from interactions of critical elements. The points that have been made are that we need to identify the individual steps in a process and do the kinetics. If we were enzymologists and we wanted to understand catalysis, we would want to have the structure of the enzyme and we would want to do kinetics. We would want to understand the steps in the process and associate a rate constant with each. Cell biology is internalizing this dictat. We are trying more and more to do structure and kinetics. When we are doing high-resolution time-lapse microscopy we are not doing descriptive biology. We have heard some cautionary notes: beware of linear explanations and interaction maps. We have heard that the systems that we are trying to understand are very complex and we need to deconstruct them by some detailed analysis. We need specific tools for this such as genetic manipulations or RNAi. Then there is a need to reconstruct all of these elements, so we need to adopt some kind of systems approach which will almost certainly involve computational modelling. Those are just some of my thoughts.

B Lane: One of the biggest problems is that we are not yet able to look at things properly in the 3D context. We still have to simplify things. Once we can start to analyse cell movement and tissue forces in three dimensions, we will need to consider other components that we haven't discussed at this meeting, such as intermediate filaments (IFs).

Borisy: We have studiously avoided IFs.

B Lane: We know that intermediate filaments are an important part of taking up the force in tissues and that defective IFs render tissues fragile. Defective keratin filaments disrupt the signalling process if the tissues are stretched mechanically or other shape changes are induced.

Borisy: Not talking about IFs was a specific omission, but the other issue you raised is a general problem: the difficulty of doing analyses in 3D. This is a serious problem for the field.

Drubin: One omission in your components list was lipid. Much of the biology we are studying here is taking place in association with lipids and lipid domains. Increasing numbers of proteins are being identified that can curve membranes or are sensitive to membrane curvature.

Sheetz: That is a mechanistic level which many of us are not yet addressing.

Borisy: Yes, it is another critical element that was omitted from the list. If there are no more comments then all that is left for me to do is to thank those involved in supporting and organizing this meeting and all of you for your participation.

Reference

Drubin DG, Kaksonen M, Toret C, Sun Y 2005 Cytoskeletal networks and pathways involved in endocytosis. In: Signalling networks in cell shape and motility. Wiley, Chichester (Novartis Found Symp 269) p 35–46

Index of contributors

Non-participating co-authors are indicated by asterisks. Entries in bold indicate papers; other entries refer to discussion contributions.

A

Alberts, A. S. 12, 26, 29, 30, 32, 33, 45, 66, 67, 71, 89, 90, 91, 123, 124, 141, 142, 191, 192, 203, **206**, 219, 220, 221, 222, 225, 226

B

Balasubramanian, M. 13, 56, 57, 69, 140
*Bass, M. D. **178**
Borisy, G. **1**, 11, 12, 14, 25, 27, 28, 29, 30, 31, 32, 33, 34, 43, 44, 45, 54, 55, 56, 58, 68, 69, 70, 71, 88, 102, 103, 104, 120, 122, 123, 124, 125, 126, 141, 142, 157, 158, 170, 171, 173, 176, 192, 201, 202, 203, 204, 220, 223, 224, 226, 227, 229
Braga, V. 101, 126, **144**, 155, 156, 157, 158, 171, 173
*Brakeman, P. **193**

C

*Cabrera-Poch, N. **106**
Cai, M. 26, 45, 57, 58, 118
Chang, F. 14, 28, 44, 57, **59**, 66, 67, 68, 69, 70, 71, 104, 119, 121, 123, 124, 139, 140, 141, 156, 172, 220, 221, 224, 226
*Chen, M. **106**

D

*Datta, A. **193**
*Deakin, N. O. **178**
*Drees, F. **159**
Drubin, D. G. 24, **35**, 43, 44, 45, 46, 56, 70, 118, 138, 169, 175, 221, 222, 229

E

*Eisenmann, K. M. **206**
*Eng, C. H. **106**

F

*Feierbach, B. **59**
Firtel, R. A. 10, 11, 14, 27, 30, 33, 44, 55, 56, 70, 71, **73**, 87, 88, 89, 90, 91, 103, 104, 121, 125, 170, 188, 225
*Fujita, Y. **144**

G

*Gassama, A. **193**
*Gomes, E. R. **106**
Gu, F. 175
Gundersen, G. G. 13, 44, 56, 104, 105, **106**, 116, 117, 118, 119, 120, 121, 122, 123, 124, 125, 126, 139, 140, 142, 172, 176, 190, 204, 221

H

Harden, N. 13, 27, 90, 139, 158, 200, 201
Hong, 170, 191
Humphries, M. J. 117, 156, 173, 174, 176, **178**, 189, 190, 191, 192, 223, 225, 226

K

Kaibuchi, K. 11, 57, 66, **92**, 101, 102, 103, 104, 105, 116, 117, 121, 123, 124, 156, 157, 171, 174, 219, 227
*Kaksonen, M. **35**
*Katz, L. **193**
*Kim, M. **193**
*Knaus, M. **47**

L

Lane, B. 119, 125, 173, 175, 176, 229
Lane, D. 12, 30, 68, 69, 70, 101, 102, 168, 169, 173, 175, 176, 225, 226
*Leroy, P. **193**
*Levin, M. **193**
Lim, L. 156, 189
*Liu, K. **193**
Luca, F. 25, 29, 203, 219

M

Manser, E. 120, 190, 191, 203
*Martin, F. **193**
*Martin, S. **59**
*Messent, A. J. **178**
*Morgan, M. R. **178**
*Morris, E. J. S. **106**
*Mostafavi-Pour, Z. **178**
Mostov, K. 13, 30, 32, 68, 69, **193**, 200, 201, 202, 203, 204

N

Nelson, W. J. 12, 13, 25, 30, 32, 56, 67, 69, 90, 101, 104, 105, 119, 120, 125, 126, 141, 157, 158, **159**, 168, 169, 170, 171, 172, 173, 174, 175, 176, 202, 204, 219, 224, 225
*Noritake, J. **92**

O

*O'Brien, L. E. **193**

P

*Peng, J. **206**
Peter, M. 14, 46, **47**, 54, 55, 56, 57, 58, 68, 69, 89, 118, 174, 175, 191, 204, 219, 220

S

*Sasaki, A. T. **73**
Schejter, E. D. 14, **127**, 138, 139, 140, 141, 142, 221
*Schmoranzer, J. **106**
Sheetz, M. 10, 14, 26, 29, 30, 32, 33, 44, 45, 57, 66, 70, 71, 87, 91, 102, 117, 120, 121, 124, 125, 140, 156, 169, 172, 176, 189, 223, 224, 225, 226, 227, 229
*Shimada, Y. **47**
*Su, T. **193**
*Sun, Y. **35**
Surana, U. 54, 67

T

Takenawa, T. **3**, 10, 11, 12, 13, 14, 15, 28, 33
*Tang, K. **193**
*Tanimizu, N. **193**
Titus, M. A. **16**, 25, 26, 27, 28, 29, 88, 119, 155, 172
*Toret, C. **35**

V

Vallee, R. 12, 13, 14, 27, 43, 103, 104, 119, 120, 121, 139, 140, 174, 204
*Verges, M. **193**

W

*Wallar, B. J. **206**
*Watanabe, T. **92**
*Wen, Y. **106**
*Wiget, P. **47**

Y

*Yamada, S. **159**
*Yamaji, T. **193**
Yap, A. 33, 155, 157, 219
*Yu, W. **193**

Subject index

A

Abi 6
Abi1 6
Abl 132
Abp1p 40
actin binding proteins 16
actin cables 60
actin filaments
  branched, Arp2/3 complex 207
  α-catenin binding 162
  IQGAP1 104
actin patches *see* cortical actin patches
actin remodelling
  current models 145, 148–149
  Rho GTPases 149–152, 179
α-actinin 162
adenomatous polyposis coli protein
    see APC
afadin 165
ajuba 165
Akt 75
amoeboid-like movement 32, 73, 74
Anc 115
APC
  cell polarization and migration 96–98
  microtubules 96, 105, 110–112, 212
aPKC 96
apoptosis 202–203
Ark1p 41–42
Arp2/3 complex
  actin filament formation 207
  actin polymerization 5
  α-catenin binding 165
  Cdc42 207
  cortical actin patches 37, 40, 41, 43
  cortical microfilaments 132–133
  formins 220
  N-WASP 5
  roles 132–133
  SCAR 133
  VCA region 5
  WASP 207
  WAVE 5, 11, 14
Ash 3–4
Ax12p 49

B

Bem1p 48, 50–51
Bem2p 48, 51
Bem3p 48, 51
Bim1 110
breast cancer 209
Bud1p 49–50
Bud2p 49
Bud5p 49
bud6p 63
Bud8p 49
Bud9p 49
Bud10p 49
bulldozer model 117–118

C

C-Cbl 4
cable car model 118
cadherin-mediated cell–cell adhesion 145
  actin 160–161
  M7a 19
  proteins linking cadherin to actin 161–165
  punctum 145
  Rac 149
  Rap1 151–152
  Rho 151
  targeting patch 197
cancer
  formins 216
  RhoC 209–210
α-catenin binding partners 161–165, 171
β-catenin 161, 166, 172
Cdc24p
  Bem1p binding 48, 50–51
  Bud1p regulation 49–50
  Cdc42p regulation 48

Cdc24p (*cont.*)
cell polarization 49, 50–51
Far1p binding 49, 50
recruitment during mating and budding 49–50
Rsr1p regulation 49–50
Cdc42
actin remodelling 150
aPKC activation 96
basolateral membrane protein trafficking 150, 152
bud formation 152
cell polarization 93
cortical actin patches 46
CRIB motifs 212
Drts 56
mDia2 212
microtubule capture 94–96
MTOC 109, 112–114, 212
N-WASP activation 4–5, 207
Par6 96
Rap1 activation 152, 156
tight junctions 150
WASP induction of Arp2/3 activity 207
Cdc42p
Bem2p and Bem3p 48, 51
Cdc24p 48
cell polarity 47–49, 51
regulation 48–49
Rga1p and Rga2p 48
CDM 135–136
CED-12 136
cell–cell adhesion 159
actin remodelling 145, 148–149
myosin VII 16–24
see also cadherin-mediated cell–cell adhesion
cell–cell contact
cadherin 145
epithelial polarization 197
interdigitated 148
cell migration
APC 96–98
IQGAP1 96–98
microtubules 59
PI3K 75
Rho GTPases 93
WAVE1/WAVE2 7–9
cell movement
PIP3 74–75
Ras 81–83
RhoA 33–34
2D and 3D 32–33
cell polarity
actin cytoskeleton 52
Akt/PKB 75
APC 96–98
Cdc24p 49, 50–51
Cdc42 93
Cdc42p 47–49, 51
cellular compass 74
destruction 57–58
ECM interactions 198
feedback-loop 83–85
IQGAP1 96–98
microtubule–actin interaction 59–66
multicellular structures 196–197
orientation 74, 196–199
Par6-aPKC complex 96
pheromone gradient 49
PIP3 74–75
Rac1 93, 196
Ras 81–83
Rho GTPases 93
site marking 54–55
tea1p 62, 63
cellular compass 74
centrosomin (Cnn) 133
CG31048 135
chemoattractants 74
chemotaxis 73–74
PI3K 75
Ras 80–81
circular waves 30
CLIP-170 94–96, 103–104
cochlear hair cells 17
cortical actin patches 36, 60
Abp1p 40
actin 40
Arp2/3 complex 37, 40, 41, 43
Cdc42 46
early patch proteins 40
Las17p 40
late patch proteins 40
movement forms 40
Pan1p 40
protein interaction network 37–39
proteins 37, 40, 41
Sla1p and Sla2p 40
Ste2p 40
cortical microfilaments 131–132
Arp2/3 complex 132–133
MTOCs 133–134, 135, 136

CRIB motifs 212
Crk 136
Crumbs (Crb) 201
cytokines
  microtubules 59
  myosins 17, 27–28

D

Dbl 50
definitions 30–31
directional sensing machinery 74
  feed-back loop 83–85
  PIP3 75
  Ras 81–83
DOCK180-like proteins 135–136
docking process 62–63, 66
Drosophila embryogenesis, actin 127–138
Drts 56
dynactin 99, 109, 113
dynamin
  Ash/Grb2-binding 4
  circular arc-like structures 12
dynein 99, 109, 113–114, 119

E

EB1 110–112, 212
EF-1α 221
Elmo 136
endocytosis 35–42
  actin 35–36 *see also* cortical actin patches
  exocytic machinery coupling 45
  receptor-mediated 35–36
  RhoB 212
  vesicle uncoating 41–42

F

F-actin
  assembly 206
  IQGAP1 cross-linkage 102
  PhDA 75
  PIP3 85, 88
  signal transduction 207
  spatially controlled assembly 207
Far1p 49, 50
fascin 166
fibrillar adhesions 179
filopodia
  definition 31
  formation 14–15
  M7 28
  N-WASP 4
  talinA 20
focal adhesions 16, 19, 179
focal complexes 179
for3, autoregulation 66
for3p
  actin cables 63
  tea1p interaction 63
  tea4p binding 64
formins
  Arp2/3 complex 220
  cancer development 216
  α-catenin binding 165
  EF-1α 221
  FH1 and FH2 domains 210
  immune cell function 216
  microtubule stabilization 109, 110
  naming 210
  regions of homology 210
  scaffolds 221
  tea1p 63–64
  tea4p 67
  WASP collaboration 216

G

G protein $\beta\gamma$ 74–75, 78
glial scars 190
Glu-microtubules 98–99
Grb2 3–4
GSK-3β 96

H

harmonin 19, 22
hepatocyte growth factor 195
HSPC300 6
HUM-6 22
hum-6 22

I

immune system, formins 216
integrins
  integrin $\alpha 4\beta 1$ 180
  integrin $\alpha 5\beta 1$ 179, 180 182, 184–185
  PKCα 186
  Rho GTPases 179
  syndecan cooperation 178–188, 191
  talin binding 16
interaction maps 37–39, 224

intermediate filaments 229
interphase microtubules 60
interphase MT organizing centres 60
invadopodia 12
IQGAP1
  actin filament stabilization 104
  β-catenin binding 166
  cell polarization and migration 96–98
  F-actin cross-linkage 102
  microtubule capture 94–96
IRSp53, WAVE2 5

K

Kar9 69, 110
Kette 6
kidney 193–194

L

lamella, definition 31
lamellipodia
  definition 31
  Rac-induced 5
  WAVE2 9
laminin, orientation of polarity 198–199
Las17p 40
leading edge, definition 31
lipids 229
lung cancer 210
lysophosphatidic acid 108–109

M

malignant melanoma 209
mariner 17
matrix metalloproteins (MMPs) 32
mDia
  microtubules 109, 110, 114, 142
  Rho GTPases 109, 211–216
  T cell signalling 216
mDia1 212
mDia2
  APC 212
  cancer 216
  Cdc42 212
  EB1 212
  microtubule stabilization 109
  MTOC 212
  RhoB 212
mDia3, cancer 216
membrane ruffling 7–9, 10, 11–12
met 195
metastasis
  RhoC 209–210
  WAVE2 12
microtubule-associated proteins (MAPs) 93
microtubule organizing centre (MTOC)
  Cdc42 109, 112–114, 212
  CDM 136
  centrosomin (Cnn) 133
  cortical microfilaments 133–134, 135, 136
  dynactin 99, 109, 113
  dynein 99, 109, 113–114, 119
  interphase 60
  mDia2 212
  myotonic dystrophy kinase-related
    Cdc42-binding kinase (MRCK) 109,
    112–113, 114–115, 120
  nuclear attachment 121
  Par6 96, 109, 114
  PKCζ 109, 114
  reorientation 93, 99, 107–110, 112–114
  sponge (*spg*) 134–135
  T cells 119–120
microtubules (MTs)
  actin interaction and cell polarity 59–66
  animal cell cytokinesis 59
  APC 96, 105, 110–112, 212
  Bim1 110
  capture, defined 122–123
  catastrophe 60–61
  Cdc42 94–96
  cell migration 59
  CLIP-170 94–96, 103–104
  directionality regulation 60
  disassembly of focal contacts 124
  dynamic instability 93, 106
  EB1 110–112, 212
  formins 109, 110
  Glu-microtubules 98–99
  interphase 60
  IQGAP1 94–96
  Kar9 69, 110
  lysophosphatidic acid 108–109
  mDia 109, 110, 114, 142
  mDia2 109
  physical tracks 141
  post-translational modification 106–107
  Rac1 94–96
  recruitment 102
  Rho GTPase stabilization 98–99, 109, 110,
    114

selectively stabilized 107–108, 110–112
stabilized 98–99, 105, 106–107, 108–110, 116–117, 118, 123–124
tea1p 62
tea2p 62
tip1p 62
+TIPs 93
microvilli 31
mod5 66, 70
mod5p 62–63
morphogenesis, genetic screens 61
muscle arms 28
muscle cell differentiation, stabilized microtubules 107
myosin II
chemotaxis 74
cytokinesis 17
PAKa 75
myosin V 17
myosin VI, β-catenin binding 172
myosin VII (M7)
cell adhesion 16–24
characterization 17
conservation 22–23
DdM7 17–18, 20–21
filopodia 28
M7/talinA mutant 25
M7a 17–19, 22
M7b 17
motor versus structural role 22–23
overexpression 25
tail domains 21
vezatin 19, 172
myosin X (M10) 25
myosins 16–17
cell adhesion 17
lack of motor function 27–28
phosphorylation 227
myotonic dystrophy kinase-related Cdc42-binding kinase (MRCK) 109, 112–113, 114–115, 120

N

N-WASP
actin polymerization 5
Arp2/3-induced actin polymerization 5
Cdc42 activation 4–5, 207
cytoskeleton 4
filopodium 4
identification 4
neural tissue 4
PIP2 5
VCA region 4–5
Nap1 6
nerve growth 189–190
neuron barrier 189–190
new end take off (NETO) 60, 68
nuclear fallout 132
nuclear migration 130–131
nucleation-promoting factors 207

P

p190RhoGAP 179, 187
PAK 90
PAKa 74, 75
Pan1p 40
Par6 96, 109, 114
PhdA 75
pheromones, cell polarity 49
PI3K 75, 76–78, 85
PI3Kγ 78
PIP2
diffusion 75
N-WASP 5
PIP3
cell movement and polarity 74–75
diffusion 75
F-actin polymerization 85, 88
GPCR 78
PTEN 75, 76–78
Ras activation 79–80
spatial localization 76
WASP 88
WAVE2 binding 9
PIR121 6
PKB 75
PKC, atypical (aPKC) 96
PKCα, integrin-specific signalling 186
PKCζ 109, 114
podosome 12, 13, 30
PREX 88
Prk1p 41–42
protein interaction network 37–39, 224
proteoglycans, neuron barriers 189–190
proteomics 224
PTEN 75, 76–78
puncta 145

R

Rac
  actin recruitment to cadherin receptors 149
  lamellipodia formation 5
Rac1
  cell polarization 93, 196
  cyst formation 196
  microtubule capture 94–96
  syndecan 4 181–182, 184–185
RacGEF 85
Rap1
  Cdc42 activation 152, 156
  cell–cell adhesion 151–152
Ras
  Ash activation 3
  cell movement and polarity 81–83
  chemotaxis 80–81
  kinetics 83, 85
  mammalian 79, 80
  PI3K activation 78, 85
  PI3Kγ 78
  PIP3 79–80
  RasGEFs 81, 83
Rga1p 48
Rga2p 48
Rho, cadherin adhesion 151
Rho GTPases
  actin remodelling 149–152, 179
  cell polarization and migration 93
  human disease 209–210
  integrin 179
  mDia 109, 211–216
  microtubule stabilization 98–99, 109, 110, 114
  molecular switches 209
RhoA
  cell movement 33–34
  Glu-microtubules 98, 99
  p190RhoGAP suppression 179, 187
  syndecan 4 181–182
RhoB
  endocytic/vesicular trafficking 212
  marker of early to late endosome transitions 219
  mDia2 212
  tumour suppression 210
RhoC 209–210
RIP3 83
Rnds 192
robustness 37
Rsr1p 49–50
ruffling 7–9, 10, 11–12

S

Sans 19
SCAR, Arp2/3 133
Scar 5
Scrib/Lgl complex 201
shaker1 17, 22
SIN 140
Sla1p 40
Sla2p 40
SOS 4
Sos 3
spectrin, α-catenin binding 164
sponge (*spg*) 134–135, 142–143
sqh 131, 132
Sra1 6
Ste2p 40
Ste3p 52–53
Ste5p 49
step, defined 223–224
stereocilia 17–19, 22
synaptojanin 4
syndecan 1 192
syndecan 2 192
syndecan 3 192
syndecan 4
  adhesion contact formation and maturation 180–181
  focal adhesions 179–180
  Rac1 181–182, 184–185
  RhoA 181–182
  Tiam1 binding 192
  ubiquitous 192
syndecans
  integrin cooperation 178–188, 191
  neuron barrier 189–190
  redundancy 192
Syne 115
syntaxin 2 194
syntaxin 3 194
syntaxin 4 194
systems approach 227

T

T cells
  mDia 216
  MTOC movement 119–120

talin 16, 19–21
talin1 26–27
talinA 20, 21
tea1 66, 70
tea1 61–62
tea1p
  bud6p 63
  cell polarity 62, 63
  docking 62–63
  for3p 63
  formin regulation 63–64
  microtubules 62
  mod5p 62–63
  tea4p 64
tea2p 62
tea4p
  for3p and tea1p binding 64
  formin activity 67
3-D ECM gels 194–195
3-D movement 32–33
tight junctions
  Cdc42 150
  gate/fence role 194
  location 197
tip1p 62
+TIPs 93
TOR complex 2 proteins 83
β-tubulin 4
tumour
  invasion/metastasis 12, 209–210
  suppression, RhoB 210

U

UNC-35 23
Usher syndrome type 1 (USH1) 19

V

VASP 26, 29
Vav 50
VCA-containing proteins 5
vezatin 19, 172
vinculin, α-catenin binding 162–164, 171

W

WASP
  Arp2/3 activity 207
  formin collaboration 216
  haematopoietic cells 4
  PIP3 binding 88
WAVE family, identification 4
WAVE1
  Abi1-Nap1-PIR121 complex 6, 13
  activation mechanism 5–6
  directed cell migration 7–9
  function 10–11
  isolation 5
  knockout mice 13
  membrane ruffling 7–9, 10, 11–12
  membranous vesicle 12
WAVE2
  Abi binding 6
  Abi1-Nap1-PIR121 complex 6, 13
  Arp2/3 binding 5, 11, 14
  directed cell migration 7–9
  function 10–11
  IRSp53 association 5
  isolation 5
  knockout mice 13
  lamellipodia formation 9
  membrane ruffling 7–9, 10, 11–12
  PIP3 binding 9
  3-D movement 32, 33
  tumour cell invasion and metastasis 12
WAVE3 5
Wiskott–Aldrich syndrome 207, 209

Z

ZO-1 164–165